Games Creditors Play

Games Creditors Play

Collecting from Overextended Consumers

Winton E. Williams

HG
3752.7
.U6
W55
1998
West

CAROLINA ACADEMIC PRESS
Durham, North Carolina

Copyright © 1998 Winton E. Williams
All Rights Reserved.

To ALICE,
for all the memories of never lost yesterdays,
and JEANNE,
who brings such joy and promise to a new day.

Library of Congress Cataloging-in-Publication Data

Williams, Winton E.
    Games creditors play : collecting from overextended consumers / Winton E. Williams.
    p.   cm.
    Includes index.
    ISBN 0-89089-991-6
    1. Collecting of accounts—United States. 2. Collection laws—United States. 3. Game theory. 4. Consumer credit—United States. I. Title.
HG3752.7.U6W55   1997
658.8'8—dc21   97-37304
                                                                           CIP

Carolina Academic Press
700 Kent Street
Durham, North Carolina 27701
Telephone (919) 489-7486
Fax (919) 493-5668
www.cap-press.com
Printed in the United States of America

Contents

Synopsis	vii
Acknowledgments	xvii
Preface	xix

1 Consumer Credit: Little Obligations of Consequence 3

2 Collection Malaise or Why Virtually No One Is Satisfied with Collection Law and Practice 11
   A. Public and Private Sector Responses to Perceived Abuses 11
   B. Creditors' Self-Inflicted Wounds in Microcosm—Kelly's Case 18
   C. Kelly's Case Generalized: A Preview of the Causes and Cures of Counterproductive Collection Practices with Ramifications for the Reform of Creditors' Remedies 22

3 Collection Practices: Competition v. Cooperation Among Creditors 29
   A. Coercion as the Product of a Race of Diligence Among Creditors 29
      (1) The Competitive Nature of Collection 29
      (2) Establishing Creditor Priority by Applying Formal Rules of Law 31
      (3) Establishing Creditor Priority by Informal or Self-Help Collection Practices 34
      (4) Beneficial Aspects of Rewarding Individual Creditor Diligence 41
      (5) The Incidence of Harmful Aspects of the Race of Diligence 42
   B. Models of Creditor Cooperation Based on Bankruptcy and Contract Law 45
      (1) Introduction 45
      (2) Developing the Bankruptcy Model: The Automatic Stay and the Treatment of Preferences and Setoffs in Bankruptcy 46
      (3) Developing the Contract Model: Composition and Extension Agreements Among Creditors 59

    C. Inadequacies in the Application of the Contract and Bankruptcy Models in Consumer Credit Cases   60
        (1) Comparing Commercial and Consumer Credit Cases   60
        (2) A Return to Kelly's Case: Factors in Addition to Unbridled Creditor Competition that May Defeat a Consumer's Attempt at an Unassisted Workout   65

4 Game-Theoretic Analysis of Creditors' Failure to Cooperate   73
    A. Modeling Creditors' Behavior as a Game of Prisoner's Dilemma   73
    B. The Transaction-Costs Barrier to a Cooperative Solution of the Creditor's Dilemma   84
        (1) The Scope of the Barrier   84
        (2) A Closer Look at Creditors' Misperceptions of the Collection Setting: Zero-Sum Games and No Contests   91

5 Resolving the Creditor's Dilemma   101
    A. Solutions Suggested by Applied Game Theory   101
        (1) Introduction   101
        (2) Tacit Reciprocity   103
        (3) Explicit Methods of Resolution and the Multi-Player Aspect of Creditor's Dilemma   111
    B. The Functions of Consumer Credit Counseling Agencies and How Those Functions Differ from Those of Bankruptcy Courts in Chapter 13 Proceedings   116

6 The Significance of the Systematic Resolution of the Creditor's Dilemma   141
    A. Preserving Creditors' Remedies While Restricting Their Counterproductive Applications   141
    B. The Problematic Basis of the Lost-Value Premise as a Sole or Principal Justification for Abolishing or Restricting Certain Creditors' Remedies   142
        (1) The Development of the Lost-Value Premise   142
        (2) A Different Perspective: The Benign Function of the Potential for Lost Value   149
        (3) Lawmakers' Attacks on Creditors' Remedies Based on the Lost-Value Premise   156
        (4) Reevaluating the Lost-Value Premise as a Basis for Proscribing Creditors' Remedies   177
    C. Preserving the Unique Role of Consumer Credit Counseling Agencies   188

Synopsis

Instead of early fulfillment of the American Dream it promises, consumer credit may spin its own dream—the nightmare of overextension. This work analyzes the forces that drive debtors' and creditors' actions when that occurs.

The book is written for anyone interested in these dynamics of consumer-debt collection. No technical background in debtor-creditor law is presumed, nor does the use of elementary principles of game theory to further the conclusions reached in the analysis of collection law and practice require that the reader be versed in game or other decision-making theory.

Because collectors often traffic in the misery of others, they seldom bask in the light of public esteem, even when the claims they collect are the result of debtor profligacy rather than debtor misfortune. Few reputable collectors resemble those villains of 19th century melodramas who fiendishly evicted starving widows on stormy evenings. Still, some of us tend to identify the contemporary bill collector with those actors bygone audiences loved to hate.

Why the work of collectors is beneficial, however, to debtors who timely honor their obligations is obvious: collectors reduce the cost and increase the availability of credit. Perhaps somewhat less obvious are the benefits that flow from the work of collectors to many other debtors—those who are motivated to confront volitional imbalances in their income and expenses while they still may be corrected without bankruptcy or other financial debacle.

But the work of collectors is not always beneficial to debtors or to their creditors. This analysis of the dynamics of the collection process explores why collectors have often insisted on immediate payment in instances in which giving debtors needed extensions for payments would appear to enhance prospects for recovery of their creditors' claims.

When innocuous requests for payments fail, collectors perforce apply increasingly compelling measures. Commonly they start with appeals to the debtor's self-esteem, move on to warnings of dire consequences resulting from loss of credit rating, escalate to threats to seize property, and, in some instances, employ their ultimate legal, if often not most economically advantageous remedy, actual seizure of some asset of the debtor. But this sequence is not invariable, and certainly seasoned collectors choose from among their available ploys those best suited to any particular case. Absent, however, the assistance of an outside agency to insure that certain conditions exist and certain measures are in place, collectors often fail to tailor their practices to accomodate extended payment plans that appear more promising than attempts at more expeditious recovery. That norm of collection practice—one of ever increasing pressure on the debtor for immediate payment of arrearages—may continue in cases in which it is counterproductive to aggregate creditor recovery.

A debtor who feels compelled to attempt payments in amounts that exceed her short-term ability will often fail. Instead of continuing her workout efforts she will effect a clandestine change of residence, simply hide non-exempt property and stonewall cost-conscious creditors till they cease their collection efforts, or seek discharge of her debts in a bankruptcy court. In any of these instances, both creditors' recoveries and the debtor's desire to honor her commitments are victims of creditors' failures to adjust their demands to conform to the means of the debtor.

A collector may erroneously reject a debtor's plea for extensions because of inadequate or misleading information concerning the debtor's needs or her resolve and ability to turn her financial affairs around. Collectors are not easily persuaded to extend payments for debtors who are already in arrears, nor in numerous cases should they be, for further delay often serves only to increase indebtedness while eroding the debtor's ability and resolve to pay. Even if the debtor's self-advocacy in favor of a workout that is promising surmounts this informational barrier, a counterintuitive assumption in many cases of consumer advocacy, another significant one remains.

In most cases of serious overextension, each creditor must believe that the debtor's other creditors likewise view the workout as promising and that they too will cooperate by giving extensions commensurate with her own. Otherwise, creditors who try to exact immediate payment of their claims may obtain preferential recoveries, while shifting the risks of an ever riskier workout to the cooperating creditor.

Obtaining the initial acquiescence of creditors in an extension plan is but one aspect of this problem of controlling harmful aspects of competition among creditors. Because complete recovery of claims in even the most promising of workouts is never a certainty, creditors who initially agree to extensions may be tempted to revert to coercive efforts to get ahead of the creditor pack, especially if the workout, as many do, runs into trouble. Insuring that this strategic behavior does not occur requires that someone monitor the conduct of creditors during the term of the workout. And as the debtor's performance must be watched and counseling given any debtors who backslide, that monitor must assume these follow-up roles as well.

Recognizing the role of creditor competition is central to understanding collection law and practice. Collection often casts creditors in the role of adversaries, for creditors of a seriously overextended debtor frequently compete with each other for their debtor's limited resources. When a debtor's finances preclude payment of all obligations, some creditors by their efforts may get more than others. Both benefit and harm may result from this contest that jurists have long described as "the creditors' race of diligence."

A system founded upon competition among creditors rewards those who are more effective at reducing their bad-debt losses, and the diligence of these creditors lowers the cost and increases the availability of credit in the many instances in which it increases aggregate creditor recovery. Either the diligent creditor collects from assets that otherwise would have been withheld from the payment of all claims, or that creditor so alters the debtor's financial course as to benefit not only herself but other creditors as well. If the race-of-diligence rule were not capable of producing such benefits, history would doubtlessly have abandoned it in its primal stages, if, indeed, it would ever have formulated it.

But failure to check creditor competition in cases in which it may diminish aggregate creditor recovery obviously increases the costs of credit and reduces its availability. The issue then is how best to switch from a system that rewards the victor in contests among creditors to one that apportions their probable enhanced recovery from participation in collective undertakings. The knotty threshold question in many cases is whether there are rehabilitation or orderly liquidation values for creditors that are worth saving by calling off the race of diligence. Bankruptcy, which imposes its automatic stay of any action by creditors to collect outside the confines of the bankruptcy court, appears to be only a part—although, doubtlessly, a very significant part—of the resolution of this issue of when creditor competition should end and creditor cooperation begin. Bankruptcy often comes too late to preserve the more fragile qualities and resources of beseiged debtors and is not particularly well suited to cases in which debtors desire to pay all or a substantial amount of their obligations because it is commonly associated with more extreme forms of debtor relief.

To the casual observer, creditor competition may appear less pronounced in the collection of consumer than commercial debt. Efforts to collect probably stop short of legal action and the use of judicial process to seize property in a much greater proportion of consumer cases than commercial ones simply because the cost of those procedures relative to the amount of the claim at stake is ordinarily much higher in consumer cases. Additionally, the prospects of recovery are frequently far lower in actions against consumers, who often have little property of real market worth and may often use exemption law to shelter any asset they possess with that worth. Corporate debtors with usually far more significant assets have no claims of exemptions. Collectors of consumer debt are understandably reluctant to risk throwing good money after bad in a futile effort to collect, so ultimate collection effort must far more often consist of talking about legal action than taking it. Whether creditors compete in or out of court, however, until bankruptcy or some collective proceeding based on agreement among creditors and the debtor is invoked, the race of diligence continues and the law rewards the victors in either judicial of informal collection contests.

Under formal rules of law, priority among creditors is based not on the time of acquisition of a claim against the debtor but on the time of acquisition of a lien or other interest, such as a cash payment, that establishes the creditor's rights against subsequent claimants to the property in contest. Law that rewards collection diligence in this manner creates a world in which creditor status is fluid. Creditors are encouraged to seek upward economic mobility by pushing for immediate payment or, failing that, acquiring secured status. The creditor persuades the debtor to grant a consensual lien or takes legal action to obtain a judicial one. A creditor with this motivation to successfully compete is disinclined to grant a debtor extensions for payment.

The race of diligence among creditors is no less pronounced when focus is shifted from priorities produced by judicial remedies, such as garnishment liens on wages, execution liens on autos and judgment liens on land, to priorities produced by informal or out-of-court collection practices. In the debtor's mind, the forum in which these contests are fought, persistence in demands for payment may elevate a creditor without a lien, and thus one of only mean status before the law, to a much higher rank. Because verbal collection contests offer such vast opportunities to establish and reorder priority in the debtor's mind, the forces they unleash on the debtor may exceed those associated with the actual seizure of property by judicial process.

Still, the obstacles to substituting a promising workout for counterproductive competitive collection practices need not deter a creditor willing to expend the resources necessary to ascertain debtor need, resolve, and ability and to obtain the continuing acquiescence of other creditors in an extension plan. But again, the size of claims in consumer cases has a pronounced effect on the process. The smaller claims indigenous to consumer credit limit the actions a prospective sponsoring creditor of a consumer workout can justify. This transaction-cost barrier to creditor-sponsored workouts is exacerbated by the complementary nature of the two factors that compose it—determining feasibility of the workout and enlisting the cooperation of other creditors. If the prospective sponsor of a workout is at least somewhat unsure of the debtor's need, willingness and ability, and therefore uncertain as to whether granting extensions will be fruitful—and such cases are virtually never free of such nagging doubt—then that creditor must recognize that other creditors harbor similar doubts that will predispose them to reject his proposal of a workout and continue their coercive collection ways.

Just as there are cases in which one creditor makes a consolidation loan paying other creditors and directly assuming all risks of the debtor's performance, there are, of course, cases in which one creditor assumes the costs of sponsoring a workout in which all creditors extend their payment terms. But difficulty in determining whether the debtor will use extensions to increase creditor recovery and vulnerability of cooperating creditors to the strategic behavior of others must necessarily substantially limit the instances of that practice in many cases where the debtor is seriously overextended.

Decision-making theory—the lessons of zero-sum games and Prisoner's Dilemma examined in this study—reinforces the conclusion reached by examining the dynamics of judicial and informal collection practices. Absent the presence of some agency designed to reduce the costs of ascertaining the probable benefits of extensions and obtaining concerted creditor cooperation in granting them, the transaction-cost barrier to workouts that appear promising may often be insurmountable.

In instances in which the benefits that will flow from all creditors cooperating with the debtor and each other in a workout are clear to every creditor, the obstacles to overcoming creditor competition that remain bear a striking parallel to those encountered by the participants in the classic game of Prisoner's Dilemma. Aggregate creditor recovery of claims is least when all creditors pressure the debtor by using coercive measures that soon cause her to abandon workout efforts, filing for bankruptcy, fleeing her creditors or stonewalling their collection efforts while hiding non-exempt assets. Aggregate recovery is greatest when all creditors cooperate in the workout by extending payment terms. But a creditor who cooperates while another employs a coercive collection strategy will recover the least amount of her claim, even less than she would have obtained if both creditors had used coercive strategies. The creditor who uses coercion against the others cooperation, scores the highest recovery of all, even more than she would have received if both creditors had cooperated in the workout, but she does so at a significant cost to aggregate creditor recovery.

While higher aggregate recovery would seem to dictate mutual creditor cooperation, that result may be obtained only if each creditor has some means of assuring herself that other creditors will cooperate in the workout. Prisoner's Dilemma teaches that in the absence of assurance of mutual cooperation, each creditor's best strategy is to employ coercive collection practices, not to cooperate with the debtor in a workout. In the prototype of Prisoner's Dilemma, stone walls, prison bars and guards keep the prisoners apart and unable to further their common good by entering into an enforceable agreement or credible mutual threats not to testify for the prosecutor against each other. In "Creditor's Dilemma" the counterpart to those physical constraints to player cooperation are the transaction costs of ascertaining debtor need, resolve, and ability and securing the cooperation of all creditors and the debtor during the term of the workout. As vivid symbols of constraint, these transaction costs pale in comparison with the devices used in the prototype of the game. Yet the bonds they place upon actors on the economic stage may be as constraining as cells to prisoners.

When creditors erroneously conclude that a debtor will not use any concerted creditor extensions given to increase their aggregate recovery, when they believe that the only issue is how distribution from a finite pool of assets devoted to payment of claims will be made among them, an additional obstacle to obtaining creditor cooperation exists. In such instances, when one creditor's gain is perceived as another's loss, the lesson of zero-sum games is that each creditor will try to outscore the other and this normally translates into each creditor trying to outdo the other's coercive collection measures. Unlike a collection case properly perceived by creditors as a Prisoner's Dilemma, there will be no incentive for creditors to try to get together and mutually extend payment terms in a contest perceived as zero-sum.

Placing creditors on the playing fields of game theory does more than reinforce conclusions regarding creditor behavior reached independently by earlier recognition of the race-of-diligence motive for coercive collection. Use of the Prisoner's Dilemma framework serves an additional and equally important function. Analysts who recognize creditors as participants in Prisoner's Dilemmas have a broader base for examining remedial measures than those analysts who fail to generalize the creditors' problem. Once the problem of securing beneficial collective action by creditors is recognized as a subset of a broader issue—the problem of obtaining collective action in all the social, economic and political contexts in which joint participation by individuals is required to achieve a common goal—the basis exists for exploring solutions to the narrower issue by application of the revelations of theorists studying the broader one.

The insights of these theorists contribute to an understanding of why a systematic resolution of the Creditor's Dilemma, apart from the traditional and significantly different one of bankruptcy proceedings, has required fashioning a unique form of private-sector instrumentality—the various non-profit Consumer Credit Counseling agencies affiliated with the National Foundation for Consumer Credit, which are primarily supported by the creditors they serve. Now widely available to debtors in this country, credit counseling agencies provide a cost-effective means of preparing workout plans when they are feasible, procuring creditor acquiescence in extending payment terms, and monitoring the actions of the debtor and creditors during the workout.

The distinctive feature of credit counseling as a private-sector remedy is that it accomplishes these functions within the severe cost constraints that distinguish most consumer collection cases from large commercial ones. By working with distressed debtors day in and day out, credit counselors hone their skills in assessing whether debtors need extensions and whether they will use them to increase the aggregate recovery of creditors. The repetitive nature of the credit counselor's work provides the specialization and division of labor upon which economies of scale are often based. By also working with the principal creditors in the community on a daily basis, the counseling agency acquires the trust and confidence of these creditors in the agency's role of monitor of the performance of debtor and creditors in a workout. In this latter function, economies of scale are most significant, for once a creditor's trust is established, the cost of securing her cooperation in subsequent cases is greatly reduced.

The affiliation of counseling agencies with the National Foundation for Consumer Credit further extends these scale economies. A newly formed agency that has received the approval of the National Foundation may trade on the goodwill of long established agencies in establishing its own reputation. And established agencies dealing with geographically remote creditors may trade on the reputations of agencies that operate in that creditor's usual trade area.

Counselor-assisted workouts differ in theory and practice from bankruptcy, the public-sector device that has long been used to discontinue counterproductive races of diligence among creditors. The difference between ordinary bankruptcy under Chapter 7 of the Bankruptcy Code and counselor-assisted workouts is marked. In ordinary bankruptcy the debtor surrenders nonexempt assets, often of little or no real market value, to creditors in exchange for freeing what is normally the debtor's principal asset, future earnings, from the claims of those creditors. Taking the opposite tack, counselor-assisted workouts commit all necessary future earnings of the debtor to paying creditor's claims in full, while leaving the debtor's nonexempt property with the debtor where it usually has its greatest worth.

While Chapter 13 of the Bankruptcy Code is designed to assist debtors who wish to use future earnings to pay part or even all of their creditors' claims, Chapter 13 proceedings are not commonly used in the same manner as are counselor-assisted workouts. Saving a detailed comparison of the two processes to subsequent analysis in this work, I note here only a few of the significant differences in the customary results of counselor-assisted workouts and Chapter 13 proceedings. Very few Chapter 13 plans propose full payment to unsecured creditors, those without liens on particular assets of the debtor; workouts sponsored by Consumer Credit Counseling agencies propose full payment to all creditors, secured and unsecured. While debtors may channel the income they are required to commit to their Chapter 13 plans to secured creditors at the expense of unsecured ones, thereby saving encumbered property that they might otherwise lose even in a Chapter 7 bankruptcy, secured creditors other than mortgagees of debtors' homes must be fully provided for in Chapter 13 plans only to the extent of the value the court places on their collateral. So neither many secured nor any unsecured creditors must be fully provided for in Chapter 13 plans. Moreover, success rates for performance of plans are higher in counselor-assisted workouts than in Chapter 13 proceedings. Thus, from the perspective of how the two processes are generally used in this country, counselor-assisted workouts complement the work of the bankruptcy courts but serve a different purpose for a different constituency than debt adjustments in Chapter 13.

No serious effort has formerly been made to normatively evaluate consumer collection law and practice from a perspective that recognizes the emergence of counseling agencies as major participants in addressing the problems of overextension. But no attempt to set the boundaries of permissible creditors' remedies and collection practices should fail to address the impact of the work of these agencies in thwarting counterproductive collection effort.

Avoiding costs to debtors that are not perceived as resulting in corresponding gains to their creditors and therefore avoiding "lost value" has long been the primary, albeit often dubious argument for banning various creditors' remedies. The principal attacks on these remedies are explored in this work. The lost-value premise has been scrutinized and found wanting in various respects by theorists, whose analyses I review, but examination of the work done by credit counselors in this study reveals an additional reason for stemming the attack on creditor remedies. Counselor-assisted workouts may be used to block the counterproductive use of remedies in instances in which bankruptcy is neither needed nor desired. Moreover, the protection afforded by the availability of credit-counseling services provides a strong impetus for reexamining laws, often ones of longstanding, that provide overly generous exemptions, which remove property from the reach of remedies that creditors could otherwise successfully employ.

Adding the credit counseling alternative to traditional means of addressing the problem of overextension in a Chapter 7 or 13 bankruptcy action gives consumers a suitable device to block counterproductive collection practices in all the various stages of overextension in which they may find themselves. They may use counselor-assisted workouts when full payment of all claims over an extended time is feasible and one of the two forms of consumer bankruptcy when it is not. The ability of debtors to employ these procedures to block creditors' use of coercive remedies undercuts the argument that some lawmakers and commentators make for banning certain of these remedies to insure that they are not used in instances in which they are perceived as inflicting greater harm on debtors than benefit to creditors.

For debtors with the ability to make payments on their obligations who fail to do so, creditors need effective remedies to reduce the costs and increase the availability of credit and to clearly signal that Americans respect the principle of honoring commitment. Recognizing that the majority of overextensions are primarily due to non-volitional expenses or interrupted income does not militate against providing creditors effective remedies to use against debtors who fail to make payment when they are able to do so. Nor does enforcement of market obligations further only materialistic values. While pursuit of the frills of the marketplace, which also accounts for much overextension, may blind us to higher cultural, intellectual and spiritual values, loss of respect for commitment to market obligation may erode these higher values as well as the fundamental tenet upon which our credit economy is based.

The argument for giving creditors more effective remedies against debtors who should not defeat their collection efforts is not meant to impugn, however, the need for bankruptcy relief in the many cases of overwhelming indebtedness that arise in this country. Whether those cases result from interrupted income or non-volitional expenses or whether financial mismanagement alone is to blame, more than loss of the impetus for future economic effort on the part of the debtor can result from failure to free remuneration for that effort from previously incurred debt. We are more than mere economic actors and need freedom from income- generating activity to pursue other aspects of our being.

In cases in which the debtor's plight is serious but not hopeless, however, these arguments in favor of bankruptcy relief must be carefully balanced with those previously made for enforcing obligation: economic policy as well as common cultural, intellectual and spiritual values also provide sound reasons for enforcing obligation. Doing what we promised another, even when it is difficult to do so, does more than just benefit the workings of markets. An even higher value is at stake here. Paying market debt is one of the most common forms of honoring the timeless religious, moral and humanistic principle of treating our neighbor as we would have our neighbor treat us. And while that principle encompasses the forgiveness of debt as well as the honoring of it, a proper balancing of these opposing aspects of the same principle requires that those whose debts are forgiven first exert good-faith efforts to meet them.

Balancing the need for bankruptcy relief with recognition of the importance of commitment to obligation might be struck by a rule discharging obligations in bankruptcy only when their enforcement would be more harmful to the debtor than beneficial to society. Indeed, a similarly highly discretionary standard examined in this work provides for dismissal of consumer cases in Chapter 7 bankruptcy for "substantial abuse". That standard may be used to dismiss cases where the debtor gives up little in nonexempt property to protect anticipated future income that is substantial in relation to the obligations the debtor is seeking to discharge. It may be too difficult a task for Congress to formulate a more specific directive to cover the myriad instances in which the issue of abuse of bankruptcy relief may arise. And it is certainly understandable that many bankruptcy judges resolve difficult cases in favor of the debtors who stand before them rather than insulate society from the injury that may result from marginal abuse of bankruptcy laws in any single case. But the effects of these individual cases mount up, and certainly greater policing of debtors who take flagrant advantage of bankruptcy is needed. Whether or not these bankruptcy reforms will materialize, giving troubled debtors who are making a dedicated effort to pay their obligations outside the bankruptcy court a means of protecting themselves and the interests of their creditors is a significant humanitarian as well as economic improvement in our collection practices.

Acknowledgments

My debt to theorists is acknowledged in the usual manner in the textual references and footnotes that follow. More difficult accounts to credit are those owing the various consumer credit counselors who first interested me in the nature of their work in casual conversations we had before I took my interest in writing on this subject seriously enough to ascribe credit to particular individuals for their contributions to my knowledge of credit counseling. Though much more concerned with analysis than detailed description of the work of credit counselors in consumer workouts, this work could not have been undertaken without the information provided by these counselors. I am indebted to Durant Abernethy, president of the National Foundation For Consumer Credit, for the data he has generously provided. Rick Tuman, Executive Director of Consumer Credit Counseling Services of Mid-Florida has also been a source of continuing assistance.

In addition to acknowledging my debt to those who do the work of consumer credit counseling agencies, full disclosure of my relationship with those agencies, whose function I evaluate in the pages that follow, requires that I note my participation, however slight, in a successful group effort to bring a consumer credit counseling office to my hometown, Gainesville, Florida. I believe my former services as a volunteer on the Gainesville Advisory Board of Consumer Credit Counseling Service of Mid-Florida, Inc. have had no perceptible effect on the detachment I have endeavored to bring to this work. The issues present in debt adjustment kindled my interests in the work of consumer credit counseling agencies and not the other way around.

Acknowledgment of greater feats of volunteerism for encouraging and assisting me in completing this work are due Peter A. Alces, Professor of Law at William and Mary, Bankruptcy Judge Thomas E. Baynes, Jr. of the Middle District of Florida, Jeffrey Davis, my colleague at the University of Florida College of Law, David Epstein, recently appointed Charles E. Tweedy, Jr. Chairholder of Law at The University of Alabama, and William Whitford, Professsor of Law at the University of Wisconsin. Larry Scott, president of Campus Federal Credit Union in Gainesville, Florida, gave me the benefit of his perspective as a consumer lender, while Scott

Collins and Todd Watson read my manuscript with the keen eyes of law students. Obviously, I am indebted in great measure to my editor, Keith Sipe, for his generous assistance in bringing this work to publication.

Wendy Cousins, David Leon, Stephanie McClain, Karen White, and Henry Sorensen were there as law students with research assistance when it was most needed as were the library staffs of the University of Florida College of Law and Hastings College of the Law of the University of California. Wendy Cousins and Karen White also provided valuable assistance in editing and suggesting revisions in format of earlier versions of this work. Helen Wheldon brought her skills in word processing to bear in preparing the final manuscript of this work, allaying my fears that the glitches couldn't be removed.

Preface

I first experienced what I have come to recognize as the pivotal problem in making small-debt collection both more effective and humane more than 30 years ago. Not long out of law school, I represented a friend in a matter that initially seemed hardly challenging. Influenced perhaps as much by the apparent simplicity of the work as by my friend's needs, I volunteered my services. My task was to obtain extensions for payment of my client's past-due obligations from his five creditors. Their debtor was awash in debt, the result of uninsured medical services and volitional overuse of credit, but made good wages. In the parlance of creditors, I proposed a "workout" in which my client would pay each of his obligations in full in 18 monthly installments. I advised his creditors that he was committed to the proposal and had no means of satisfying their claims other than from that part of his wages not required for subsistence, the part that he would use to fund the workout. He had neither property of real market value that could be sold to pay creditors nor friends or relatives to whom he could turn for financial aid.

I considerably underestimated the difficulty of obtaining his creditors' acquiescence in the proposed workout. They continued to hound my client demanding immediate payment of their claims. Their demands were backed by threats of suits, which led, in their more vivid descriptions, to garnishment of wages, loss of card-carrying membership in the credit society, and social censure that would not be limited to his life on this earth.

I had carefully prepared a budget with my client and his wife, taken into account not only his wages but also hers from a part-time job, and committed every dollar over those necessary for the support of their family, which included two children, to payment of the creditors' claims. The amount allocated to funding the workout exceeded the amount the creditors could have obtained by garnishment of all wages not protected by exemption laws. Risk of failure of the workout was shared equally as each creditor received monthly its pro-rata share of the amount the debtor had contributed to fund the workout.

When my client continued to receive dunning letters and phone calls threatening suit after his creditors received written copies of the workout proposal along with checks on my trust account for their first payments, I called his creditors again. Most gave my arguments short shrift, declaring, as though they were invoking a principle of natural law, that they had no authority to alter the events inextricably set in motion by my client's "willful failure to pay his just obligations." One unrelenting creditor asked incredulously: "You want us to give him more time to do what he already isn't doing?" None of the creditors, as I recall, showed the least bit of appreciation or even civility for what I believed were my efforts on their behalf as well as those of my client, and some were downright hostile.

Only by suggesting the possibility of a bankruptcy filing if any of the creditors did sue—a ploy that was made less effective by my client's employer's then stated policy (now unlawful) of firing employees who "took" bankruptcy—did I eventually succeed in securing the begrudging acquiescence of the five creditors. But some of them ceased their incessant demands for immediate payment of their entire claims only after they had received several monthly payments.

Significantly, the workout succeeded. Creditors received payment in full except for some losses due to the time value of their money, a matter of lost interest that seems trivial when compared with the losses they would have sustained had a garnishment action by any one of them caused my client's employer to fire him. Service of a garnishment writ on an employer requires that employer to pay the nonexempt portion of its employee's wages into the registry of a court, and in the mid-1960's some employers, including my client's, sometimes reacted to this annoyance by terminating the employment of the employee whose obligation had given rise to the garnishment, another action that would be illegal today.

Now, after some three decades of teaching, researching and consulting has supplemented the lessons learned from representing creditors and debtors in those early years of practice, I have a better understanding of the dynamics of debt collection. There are forces driving the law and practice of consumer-debt collection that dispose creditors to use coercive collection measures in instances when cooperation with their debtor in a workout would appear to enhance their prospects of recovery. These are instances in which creditors' failure to extend payment terms destroys the debtor's economic viability, and the debtor files for bankruptcy, absconds with all property of any value or simply stonewalls creditors who are thwarted from using legal process to collect by ineffective remedies or transaction-cost barriers to the use of effective ones.

When innocuous requests for past-due payments fail, creditors apply increasingly coercive measures in their attempt to effect recovery of their claims. These measures may culminate in seizure of debtors' assets under legal process, but far more often they proceed on the threat of that process, accompanying embarrassment, and loss of credit rating and self-esteem to bring forth payment.

Another factor augments this tendency of creditors to increase pressure on debtors when less stringent measures fail. Creditors know that debtors who fail to respond to past-due notices, especially those that arise from discretionary, undisputed obligations, are apt to be in arrears on more than one debt. Competition among creditors for a debtor's limited resources induces each creditor to try to outdo the coercive measures employed by the debtor's other creditors. In some instances, however, this contest is not won by the creditor whose collector talks toughest about the use of judicial process and other unpleasant consequences of nonpayment. Some seasoned collectors sympathize, at least outwardly, with the plight of those debtors who, they feel, will respond more readily to sympathy than threats. Their mode of collection is still coercive, however, when the concern they manifest for the debtor is designed to obtain priority in payment over other creditors and not to give the debtor needed extensions for making payment.

Recognizing creditors' need to use coercive collection measures to turn recalcitrant debtors around and to successfully compete with other creditors does not explain, however, why creditors often fail to switch to the cooperative mode of sharing what their debtor can afford to pay in a workout when that method of collection affords them the probability of greater recovery. Is the germ of cooperative effort for mutual benefit so missing from the psyche of creditors that they must be taught to share their entrees in a Chinese restaurant? Something other than a failure of creditors to subscribe to the principle of maximizing utility must explain their failure to work with the debtor and each other in consumer workouts that appear advantageous.

In exploring why beneficial cooperative action is liable not to be taken by creditors who are unassisted by a public or private agency designed to effect that purpose, this work pays particular heed to the role of deficiencies in creditors' knowledge. Creditors have reason to be skeptical of information supplied by debtors regarding their need for extensions and their intent, commitment and ability to perform a workout. Giving requested extensions to debtors may often exacerbate the problem of repayment. Some people are credit addicts and cease buying more than they can afford only when the bill collector successfully dramatizes the error of their profligacy. Information is not costless, and creditors operating under cost constraints have difficulty in distinguishing cases in which extensions will increase recovery from cases in which extensions are counterproductive.

In addition to these informational deficits regarding debtors' need for extensions and their ability to perform a workout, I examine a further major obstacle to implementing workouts, one that imposes informational, negotiating and monitoring costs that must be rigidly contained. Any creditor engaged in a contest with other creditors for resources and debtor resolve that may prove inadequate to pay all claims can ill afford to grant concessions to the debtor unless other creditors do likewise. Therefore, before cooperating in a workout, each creditor needs to insure that what jurists have long termed "the race of diligence among creditors" has ended.

Obviously, none of these impediments are insurmountable in a world without cost constraints, but, where economic considerations restrict choices, a creditor may not be able to tailor collection methods to the particular needs of the case. And creditors encounter a transaction-cost barrier much sooner in collecting consumer-credit claims than commercial ones due to the relative size of the two types of claim.

Concerns about the misuse of collection methods reached by examining the principles underlying collection law and practice are reinforced by simple applications of elementary game theory, which assume no prior knowledge on the part of the reader. Both Prisoner's Dilemma and zero-sum games are used to further explore the difficulties of getting creditors to cooperate absent some intervening agency in even the more promising workouts proposed by consumer debtors.

Further applications of elementary game theory explain how the emergence of an agency in the private sector, Consumer Credit Counseling Services, which has been supported by creditors and in some instances community giving through organizations such as the United Way, has provided creditors and debtors far greater opportunities to engage in advantageous workouts. These counseling services determine a debtor's need for extensions, assess the debtor's commitment and ability to successfully complete a workout based on those extensions, convince creditors of the merits of the workouts they propose, and, finally, administer the workout by receiving payments from the debtor and disbursing them to creditors. And Consumer Credit Counseling agencies perform these services within the stringent cost constraints that the relatively small size of consumer-credit claims impose on collection procedures. These counselor-assisted workouts differ in theory and practice from bankruptcy, the public-sector device that has long been used to discontinue counterproductive races of diligence among creditors.

There are conspicuous differences in ordinary bankruptcy under Chapter 7 of the Bankruptcy Code and couselor-assisted workouts. In ordinary bankruptcy the debtor trades his non-exempt assets for the release of his future earnings from the claims of pre-bankruptcy creditors. Market forces would rarely ever produce such a trade. The debtor's non-exempt assets usually have little or no market value and thus seldom provide real benefit to creditors. The debtor's future earning ability, virtually always his most valuable asset by far, is precisely what his creditors had looked to for payment. Counselor-assisted workouts retain this pre-transaction focus and commit all necessary future earnings of the debtor over those required for subsistence to payment of creditors' claims, while leaving the debtor's possessions where they usually have their greatest value, with the debtor.

Although Chapter 13 of the Bankruptcy Code was specifically designed to assist debtors' use of future earnings to pay creditors' claims, subsequent examination will show that counselor-assisted workouts and Chapter 13 proceedings ordinarily differ significantly in use and result. Notwithstanding Chapter 13's provisions protecting creditors, debtors may contribute significantly less of their future earnings to payment of their debts in that bankruptcy proceeding than they do in counselor-assisted workouts. Moreover, much of debtors' earnings contributed to Chapter 13 plans may be allocated to paying the claims of their secured creditors. These creditors have liens, which are interests in particular assets of the debtor that secure the creditors' claims. By directing payment to these secured creditors, as they are permitted to do under terms of the Chapter 13 plan, debtors may retain property encumbered by liens that they would otherwise lose even in Chapter 7 bankruptcies. The result is that unsecured creditors, those who do not have a lien on some asset of their debtor, often get little or no more in Chapter 13 than they do in Chapter 7, where, as previously noted, their recovery is frequently scant or nonexistent. The goal of counselor-assisted workouts is full payment of all creditors' claims in installments made during the term of the extension obtained by the credit counselor.

By offering an alternative to coercive methods of collecting debt from overextended consumers in addition to bankruptcy, counselor-assisted debt adjustment sheds new light on the solution to an age-old problem. We need to put teeth in the law of monetary obligation, or else why voice that law. But our enforcement mechanisms must respect the personal freedom of the individual as debtor no less so than we respect the personal freedom of the individual in other capacities.

In principle, the laws of the various states of this country provide creditors who satisfy procedural safeguards, which impose necessary costs in time and money, the right to seize a defaulting debtor's property except that which is protected by exemption laws, which differ widely from state to state. But overriding federal law, the discharge provisions of bankruptcy, accords the right of people to enjoy the present benefits of their labors higher status than it does compulsory commitment of those benefits to repayment of debt. The cost to the debtor of sheltering the fruits of present effort from previously incurred obligations imposes no charge on those efforts in a Chapter 7 bankruptcy proceeding where costs to the debtor are limited to loss of non-exempt assets. Bankruptcy relief, which gives the debtor a fresh start, may be based upon the need to retain economic incentives to keep overextended debtors productive. But that relief is also justified by the recognition that prolonged loss of present use of wages, like the peonage and indentured servitude of the past, imposes too high a price in personal freedom for any economic benefits it may bestow on society.

Bankruptcy, however, is often not invoked or not invoked until after some debtors have been subjected to collection processes that are unduly harmful to them and unproductive or even counterproduc-tive to their creditors. The adverse results in these cases have caused lawmakers to place various restrictions on remedies creditors may use outside the arena of bankruptcy to subject debtors' property to their claims. These restrictions on remedies have grown in recent years, and creditors often have no effective means for effecting recovery in cases in which debtors should clearly be held accountable for their obligations. What is needed is a means of limiting the use of creditors' remedies in cases where they are harmful without abolishing or unduly circumscribing the use of those remedies in other instances. Consumer Credit Counseling agencies significantly assist creditors in demarcating between productive and counterproductive use of their coercive collection remedies.

Counselor-assisted workouts provide an alternative for debtors who do not need the protection of bankruptcy but who are in need of adjusting their obligations by obtaining extensions from their creditors. Because counselor-assisted workouts, which are now available to most Americans, give qualifying debtors a method of remedying their financial problems without having to choose between the sting of coercive collection or the stigma of an unwanted bankruptcy, the case for restricting creditors' remedies to prevent their deleterious effects in some instances is considerably diminished. Overextended debtors now have the power to block counterproductive collection methods by using some form of bankruptcy when payment of all claims over even an extended time is beyond the pale and by using counselor-assisted workouts when such payment is feasible. For debtors not paying their obligations who fall in neither category because they are able with adjustments in budget to make timely payment, creditors need effective remedies to reduce the costs of credit and to signal clearly that our society honors the law of commitment in the marketplace.

Not everyone, of course, shares my view of the benefits of Consumer Credit Counseling agencies. The reaction of a friend who has taught, practiced, and researched insolvency law for many years to a cursory description of this work expresses this contrary view. In his opinion, workouts through consumer credit counseling agencies often impose more sacrifice than society ought to exact for our necessary purchases, or even our foolish excesses, in the marketplace. He would have many of the candidates for counselor-assisted workouts resort instead to bankruptcy and probably to Chapter 7 to immediately shelter their future earnings, unless Chapter 13 was needed to assist debtors in retaining property they valued that would otherwise be lost to creditors.

The reasoning of some who think this way is that even if the excessive debt is volitional, resulting from expenditures or loss of income that the debtor could have reasonably prevented, it is induced by a corrupt market that beguiles innocent consumers by creating not only the product but also the demand for the product. To those who are well versed in this language of victimization and who accept its tenets, the law should be replete with measures that give the consumer easy escape from market predators, especially those who capitalize on what is often no more than the illusion that paying tomorrow will be easier than paying today.

I am of a different mind, and it is reflected in those parts of this work where value judgments must be made. When debtors' loss of self-respect and ability to obtain future credit are considered, those who would provide them with easy escape from their obligations may be seen as not having a lock on compassion. The freedom to make right choices in the market must necessarily impose some responsibility for making wrong ones, and the free market and an effective credit system to enhance it are values worth embracing. I do not view as misguided those who labor mightily to pay their obligations when discharge in bankruptcy is a Federal courthouse away. To the contrary, the economic benefits of the sound use of credit for individuals, businesses and other entities, including nations, could not exist without most debtors being firmly committed to honoring their obligations.

Games Creditors Play

1

Consumer Credit: Little Obligations of Consequence

Obligation cannot exist absent some measure of personal commitment on the part of the obligor or some enforcement mechanism provided by custom or law. In a democratic society, these individual, social and legal agencies typically reinforce one another. But they often exert greater effort on behalf of the performance of some types of obligations than others. And various individuals and groups often differ widely in their commitment to discharging similar obligations. These diverse attitudes toward the extent of our duty to perform obligations are no doubt a function of different measures of collective respect for various obligations and different measures of individual respect for similar ones.

This work explores individual, social and legal responses to one of the most common forms of obligation—credit extended to individuals for obtaining the necessities and frills of the marketplace in a highly interdependent society. By sheer quantitative measure then, we explore an important facet of the enforcement of obligation. True, how we honor or fail to honor an obligation to a bank or merchant is not ordinarily as important a social issue as how we treat a spouse or child, or whether we use force to control our neighbors or their property. Still, how we treat one type of obligation influences how we treat others, and society's expectations concerning credit obligations help shape far more aspects of our life than just our experiences in the marketplace.

To explore, however, the immediate significance of our responses to credit obligations, a subject that is critically important in its own right, the function of credit in a market economy should be traced. For purposes of this study, an outline of credit's contribution to economic development through credit's contribution to trade requires far fewer lines than the outline's chronological beginning in prehistory may suggest.

The evolution of our primal ancestors from hand-to-mouth foragers to hunters and gatherers who stored and shared their food demanded a degree of social cooperation not found in other primates. And the division of labor, social cohesion and pattern of sharing developed over at least a million years of hunting animals and gathering plants for delayed consumption made possible the agricultural revolution of 10,000 years ago, the industrial revolution of the 19th century and the technological one of the 20th.[1]

Some sharing consists simply of helping others without thought of recompense. Altruism is in tune with our deepest beliefs and values and accounts for a considerable amount of human effort. But because an individual's or group's resources are virtually always limited relative to their material needs and desires, there are limits to the effort most of us will exert to produce goods or services for others when that effort may confer psychic or spiritual but no material benefits on ourselves.

What we and our primal ancestors needed to move from a hand-to-mouth existence to an age of relative affluence was something other than some of us working gratuitously for the benefit of the rest of us, regardless of how ennobling that may be. What we needed for those great strides forward was, first, an extension of the primordial family model of working together to reap the benefits of additional division of labor in surmounting the problem of day-to-day survival. Once that was done, we could then go on to construct a capital base of knowledge and material resources upon which future cooperative effort would produce, at least in theory, an ever improving quality of life fueled by ever increasing gains in the capital base.

How we enlist individuals in the work of humankind to better ourselves is no mystery. The prospect of individual compensation is a strong impetus. The larger share of society's wealth a person is promised for higher productivity includes greater social recognition as well as the ability to consume more goods and services.

The reward for productive effort, as well as the organization of production itself, may be determined by the state. But absent direct state control of the economy—a method for securing cooperation that has fallen increasingly into disrepute in this last decade of the 20th century—reciprocity, which in market economies is manifest as trade or exchange, must necessarily furnish the incentive for most of the production that we share. Throughout human existence, trade has induced cooperation by invoking its own kind of magic.

The flint knapper of 30,000 years ago who traded spear points for a cave-bear skin to use as a ceremonial robe for his shaman doubtlessly valued the bear skin more highly than the spear points, while his trading partner took the opposite view of the relative worth of the goods. Was one party mistaken or is it possible that each of them got more value than he

---

1. *See* RICHARD E. LEAKEY & ROGER LEWIN, PEOPLE OF THE LAKE 126–59 (1978); RICHARD E. LEAKEY & ROGER LEWIN, ORIGINS 148–49 (1977).

gave? To put the question in the language of economics, does one party's increased utility not necessarily come at the expense of the other? One observer of this phenomenon has suggested that the use of trade to capitalize on differences in relative utilities among individuals may be described as the creation of wealth out of thin air.[2]

But wealth was created for the Upper Paleolithic flint knapper and bear hunter only if each got what he reasonably thought he was going to get when the bargain was struck. And the law of modern times, and probably that of prehistory as well, has had to concern itself mightily with two commonly encountered impediments to trade producing the increased utility the parties envisioned.

If one of the spear points was defectively cut and broke on impact with the next cave bear the hunter encountered, or if the bear skin was improperly cured and failed to yield the years of trouble-free service the flint knapper and his shaman, obviously without power to foretell the future in this regard, justifiably expected, the trade was flawed. In either instance, the injured party probably did not increase his utility at all and certainly did not do so to the extent he expected.

The complaint of the hunter of cave bears or more likely that of his heirs, for his was an especially dangerous occupation to practice with faulty tools, would lie today in breach of warranty under contract law, unless the contract was devoid of any applicable warranty of quality.[3] But even if no warranty was given, an action could now be brought in strict liability in tort. The state imposes a quality standard for safety in the sale of an unreasonably dangerous product that is not dependent upon the agreement of the parties to the sale or proof of the manufacturer's negligence.[4] The flint knapper's injuries, being limited to disappointed economic expectations resulting from the need to replace the shaman's flawed bear skin before the next high holy day, must find a basis for recovery in modern times in an express or implied warranty in the contract of sale.[5] While defective product quality mars enough trades to command the expenditure of significant legal and emotional resources and therefore justifies its inclusion in this sketch of the role of trade in organizing cooperative effort, the concern of this essay is solely with the second of the two principal impediments to the wealth-producing powers of trade.

When contemporaneous exchange of the consideration each party is to provide the other in a trade does not occur, one of the parties to that transaction must incur the disadvantage of being the first to perform. Obvi-

---

2. ARTHUR A. LEFF, SWINDLING AND SELLING 10 (1976).
3. U.C.C. § 2-316 (1990).
4. *See* RESTATEMENT (SECOND) OF TORTS § 402A (1965).
5. U.C.C. §§ 2-313 – 2-315 (1990).

ously, one disadvantage of being the first performer is that that party receives her promised benefits later than she would in a contemporaneous exchange, and any delay in receiving her *quid pro quo* is disadvantageous for the simple reason that goods, services, and that in which we express their worth, money, have value over time. One would rather have a dollar today than that same dollar a year from now and will pay the rate of interest charged by a lender to do so. This aspect of first-performer disadvantage is not an impediment to trade but actually an opportunity to trade for something additional—credit—for which a separate charge is exacted—interest.

There is, however, a disadvantage to being the first party to perform one's contractual duties that may significantly impede trade's flow of resources from lower to higher valued uses. The party who is to tender the later performance may simply fail to do so. While the interest charge for delayed performance will include, in addition to the time-value of money, a component to cover anticipated losses from default, the party to whom performance is due, the creditor, is left with the risk that she may incur a higher loss rate than the amount the market or government through regulation has permitted her to charge. Commerce responds to this risk inherent in the giving of credit in various ways.

First, the parties to a contract may simply avoid an extension of credit by concurrent performance of their duties. The Uniform Commercial Code recognizes that in a contract for the sale of goods, neither buyer nor seller should assume a credit risk in the absence of an agreement to do so, which may be express or implied through trade usage, course of dealing or course of performance. In the absence of such overriding agreement, however, the Uniform Commercial Code's fail-safe rule produces the constructive concurrent condition of exchange, which we call a cash sale. The buyer's duty to tender payment for the goods is conditioned upon the seller's tender of the goods, and the seller's duty to tender the goods is conditioned upon the buyer's tender of payment.[6]

Contracts providing for concurrent conditions of exchange, however, are often not convenient, sometimes not practicable. In a service contract, where the work of the service provider cannot be performed instantly, it is hardly practicable for the recipient of the service, the customer, to make numerous payments to cover each stage of completion as the work progresses. Either the service provider must give credit to the customer by completing the work or some agreed portion thereof before the receipt of payment, or the customer must give credit to the service provider by paying in advance of the performance of the work.

---

6. U.C.C. §§ 2-507(1), 2-511(1) (1990).

Faced with the obvious need for a credit transaction, the parties to a service contract have customarily placed the credit risk on the party best able to prevent or minimize loss from nonperformance by the other. As the service provider's work is usually the repair or improvement of goods of the customer that are perforce in the service provider's possession, an assist from the law, the giving of a lien on these goods not dependent on consent or judicial action but arising solely by operation of law from the status of the parties, puts the service provider in the position of being the party best able to bear the credit risk. Even if the service provider does not gain possession of property of the other contracting party, such as when the work is performed on a customer's realty, or when the service provider works on an employer's goods at the employer's place of business, there is an obvious choice of property upon which a lien may be fixed: the realty or goods upon which the work was performed. Were the recipient of service to pay before the service was performed, she would have no property of the service provider in her possession. Nor is there necessarily any obvious type of property owned by all service providers to which a statute or other rule of law could look to prescribe a non-possessory lien.

Structuring transactions so as not to use credit can pose problems even in the sale of goods, where the performance obligations of the parties lend themselves to concurrent performance—the exchange of money and goods. Where buyer and seller are separated by distance so that a face-to-face transaction is not practicable, the parties must incur additional costs for the services the carrier performs in collecting the price from the buyer in a cash-on-delivery sale.

Where the parties to a substantial commercial transaction also operate in different geographic markets and wish to avoid credit risk, the seller will probably ship the goods to the buyer under a negotiable bill of lading, the possession of which controls the right to receive the goods from the carrier. The buyer may obtain possession of the negotiable bill of lading when it is tendered by a bank in her area only by paying an accompanying sight draft drawn upon her by the seller. In this manner, the seller is assured that the buyer cannot obtain possession of the goods until the price is paid, and the buyer is assured that upon payment she will receive a negotiable bill of lading that controls the right to receive the goods from the carrier.[7]

Although the seller will not suffer loss of possession of the goods to the buyer if the buyer fails to pay the sight draft, that breach by the buyer may

---

7. Robert J. Nordstrom, John E. Murray & Albert J. Clovis, Problems and Materials on Sales 312–23 (1982).

damage the seller in other ways. The seller's freight costs may add no value to the goods in the form of place utility. To the contrary, the goods may be disadvantageously located for purposes of further sale or in danger of spoilage or other type of depreciation. To minimize the risk of breach, the seller may insist that the buyer obtain a commercial letter of credit from her bank, whereby that bank promises to pay a sight draft for the purchase price of the goods when the seller presents the bank the negotiable bill of lading. Because the bank is far less likely to be insolvent than the buyer, and because the bank will not have the same incentive to escape its obligation in a falling market as the buyer, there is less risk of nonpayment by the bank than the buyer.[8]

These procedures that enable buyer and seller to avoid credit risks even when they are separated by distance impose additional costs on doing business that weigh in favor of the use of credit. To approach credit, however, as a device that is used when the nature of the transaction or geographic separation of parties place burdens on concurrent exchange is to project a cameo role for credit while ignoring the leading role it plays in our market economy.

There are myriad credit transactions in this country because debtors believe the benefits of credit exceed its costs, while creditors believe the interest they receive for their capital exceeds the loss of their ability to put that capital to other uses. An additional benefit exists for credit extenders who are also sellers of goods. They reap the profit on those sales they would not have made without extending credit to buyers. The use of credit by sellers to increase their sales volume is probably as old a sales-promotion device as trade itself, for any young man getting started in the business of hunting cave bears would require spears tipped with the finest flint points before he could hope to deliver his first bear skin in trade to the flint knapper.

The market for credit, like the market for goods, acts upon the different relative utilities of the parties to the transaction to create wealth out of thin air. In today's economy, credit enables consumers to obtain housing and durable goods much sooner than they otherwise could. And what a convenience the credit card offers as a cash substitute and flexible means of balancing monthly disbursements with income. Far from being something that is grudgingly given by sellers only when a sale may be made in no other way, credit is aggressively marketed throughout our country, frequently by financial institutions and finance companies that do not share in the profit from the sale of the goods or services that give rise to the extension of credit.

---

8. JAMES J. WHITE & ROBERT S. SUMMERS, UNIFORM COMMERCIAL CODE 806–09 (3d ed. 1988).

In light of this elementary market observation, the use of cash on delivery sales, negotiable documents of title and commercial letters of credit can now be seen not as devices to permit contracting parties to continue a pattern of cash sales, for in most markets credit sales are far more common than cash ones. Instead, these devices that permit concurrent exchange are used for aberrant transactions, either ones in which the buyer has a poor or unknown credit record or ones in which the effectiveness of the judicial system that would be used to enforce the obligation against a buyer in breach is questionable or unknown.

The use of devices to foreclose the need to extend credit and to protect creditors in instances, such as service contracts, when credit necessarily must be extended, do illustrate, however, the constant concern creditors must give to the risk of default. Obviously, creditors who wish to remain in business will carefully evaluate an applicant's capacity to repay the requested loan and the applicants's record in discharging previous obligations, a quality known in the credit industry as character. Evaluations combining capacity and character produce a measure known as creditworthiness, which, when coupled with a prudent use of any available collateral to ease the burden of those defaults that do occur, afford the creditor all the pre-default protection possible in an uncertain world.

But foretelling the future of the debtor's fortunes, the strength of her character and the condition and value of collateral at the time a default may occur is highly inexact work. Even low-risk creditors see good loans go bad, and, when that occurs, what creditors and debtors need are flexible measures for dealing with the various types of recovery problems they may face.

Post-default procedures are obviously important to creditors and debtors who become embroiled in the collection process. But the ways we enforce or fail to enforce monetary obligations have greater consequences than their impact on the fortunes of parties to credit contracts in default. Creditors use the judicial remedies of collection law and, far more frequently, the self-help practices that the law permits, to cut their losses and therefore the costs that they must generally pass on to debtors who pay. Absent a high rate of payment of creditors' claims, affordable credit could not exist.

Recognizing the contribution of collection law and practice to credit, credit's contribution to trade, and trade's contribution to productivity link the subject of this work to the basics of our economic life. Debt collection, however, is more than an economic issue. Probably nothing shapes the way a society sees itself more fundamentally than the way it enforces or fails to enforce the performance of obligations. While there are more important social and familial obligations than the payment of monetary obligations incurred in the marketplace, consumer credit represents the source of a great deal of our learning about our responsibilities to others. Thus, the way we respond to paying our ordinary bills and the signals the law gives of society's expectations in that regard lie at the roots of human relations. Unfortunately, the law has brought considerably more sophistication to bear in establishing rules for determining liability than it has in fashioning mechanisms for enforcing or discharging the liability it has established.

2

Collection Malaise or Why Virtually No One Is Satisfied with Collection Law and Practice

A. Public and Private Sector Responses to Perceived Abuses

Deferred gratification finds little favor with consumers, but credit may produce the tensions of overextension instead of the early fulfillment of the American Dream it promises. The law responds to credit gone awry by providing rules for the collection of claims, which are ultimately administered by the courts. But collection law garners few plaudits from those who use and are subjected to it and little critical acclaim from its practitioners and analysts. Now emeritus Harvard Professor Vern Countryman concluded more than 20 years ago that

> [t]he field of debtor and creditor law is rough terrain indeed. To the extent that the law is supplied by the states [and except for the federal law of bankruptcy, most all debtor and creditor law is state law] it consists of a hodgepodge of remedies, mostly developed in England years before the Revolution and adopted and maintained here without systematic re-evaluation.[1]

In the words of another respected commentator of longstanding in this field, Professor Stefan A. Riesenfeld of that treasure-trove sometimes euphemistically styled "the Over 65 Club" at Hastings, the law of creditors' postjudgment remedies "are often cumbersome, clumsy, inequitable and overly technical."[2]

---

1. VERN COUNTRYMAN, CASES AND MATERIALS ON DEBTOR AND CREDITOR lxiii (2d ed. 1974).
2. Stefan A. Riesenfeld, *Collection of Money Judgments in American Law—A Historical Inventory and a Prospectus*, 42 IOWA L. REV. 155, 181–82 (1957).

While my observation that collection law and practice have few if any fans is admittedly impressionistic, it is based, at least in part, on the conclusions of these and other impartial, analytical observers. Another form of pervasive criticism among academics of the law governing debtors and creditors has less significant roots in scholarly analysis. One indication of the somewhat unsavory reputation of "collection law," a term commonly used by practitioners, is that no American law school of which I am aware denominates the non-bankruptcy component of its course in the law of enforcement of monetary obligations in such fashion. Rather, law schools refer to it by such names as "Creditors' Remedies, "Creditors' Rights," "Creditor-Debtor Law," "Debtor-Creditor Law" (the difference in the latter two reflecting a myopic concern with top billing), or, the more palatable but unwieldy term, " Debtors' Protection and Creditors' Remedies," which, of course, is also subject to inverse-order arguments.

Some of my colleagues in law-school teaching are prone to wince when on occasion I confess to teaching collection law. But, indeed, this nomenclature issue may be something more than a euphemistic enterprise directed at an admittedly pathological area of the law, for there is sometimes something in a name for some of us. A student of mine once asked, in all seriousness, if the law school offered a companion course to mine, which was then styled "Creditors' Rights," for those who wanted to represent debtors. I hope his question was asked, as I recall it was, on the first day of class and that my presentation of materials had not raised the student's concern.

An examination of the serious commentary on collection law reveals that it is criticized both for its strengths and weaknesses in subjecting debtors' property to the payment of creditors' claims.

Some observers feel that anemic creditors' remedies contribute to unjustifiable losses that are usually passed on to debtors who pay.[3] Bankruptcy, a symptom of the many factors that cause loan loss and also a cause of those losses in instances in which its relief is abused, will reach more than a million cases in this country in 1996, an ominous new high in fil-

---

3. *See generally,* e.g., HOMER KRIPKE, CONSUMER CREDIT 295–300 (1970) (the classic argument from microeconomics—restricting creditors' remedies may increase the costs and limit the availability of credit—made by a lawyer and law teacher with considerable hands-on experience in the consumer-finance industry); ALAN SCHWARTZ & ROBERT E. SCOTT, COMMERCIAL TRANSACTIONS 948–50 (2d ed. 1991) (the costs to consumers of prohibiting creditors' remedies exceeds the benefits they derive from such action); Homer Kripke, *Consumer Credit Regulation: A Creditor-Oriented Viewpoint,* 68 COLUM. L. REV. 445, 478–86 (1968).

ings.[4] Reports of high delinquency rates and large bad-debt losses in some of the credit industry's principal lending programs have been common throughout this decade.[5] Moreover, media coverage of such events as the precarious financial condition of many banks,[6] high default rates on government guaranteed loans[7] and large losses booked by lenders to less developed countries[8] have focused public attention as seldom before on the adverse effects of systematic loan losses. The public's increased awareness that costs from bad debt losses fall on debtors who pay and sometimes on taxpayers must have influenced the angry public reaction to the report that more than a few members of Congress in 1991 wrote bad checks on the House bank and failed to pay bills owed the House restaurant.[9] While systematic loan losses obviously cannot be attributed solely to inadequacies in creditors' remedies for many complex socio-economic factors are at work here, the efficacy of permissible collection practices is a factor worthy of serious consideration in any effort to preserve and enhance a viable credit economy.[10]

---

4. Michelle Singletary, *A New Breed of Debtor Shocks Credit Card Issuers: As Delinquency Rates Hit Record Highs, It's Affluent, Low-Risk Customers Who Are Declaring They Can't Pay*, WASH. POST, Sept. 18, 1996, at F01.

5. *See, e.g., id.* (rising delinquency rates for credit cards and eight types of "closed-end installment loans and increased bank credit card losses); L.A. TIMES, Dec. 13, 1990, at A1; WASH. POST, Dec. 21, 1990, at B1 (losses on credit cards); N.Y. TIMES, July 17, 1991, at A1 (losses on real estate loans).

6. *See, e.g.*, WASH. POST, Oct. 27, 1991, at H1.

7. *See, e.g.*, CHRISTIAN SCI. MONITOR, Aug. 16, 1990, at US7 (student loans); WASH. POST, Nov. 24, 1990, at E1 (V.A. loans).

8. *See, e.g.*, Editorial, *Passing the Plate for Kuwait*, WASH. POST, Oct. 26, 1991, at A20.

9. For an example of this reaction, see Editorial, *Kitegate Spills Over*, WALL ST. J., Oct. 4, 1991, at A14.

10. *See, e.g.*, DOUGLAS G. BAIRD & THOMAS H. JACKSON, CASES, PROBLEMS, AND MATERIALS ON BANKRUPTCY 3 (2d ed. 1990) ("Commerce as we know it can thrive only if the obligation of the debtor is legally enforceable: If a debtor does not pay what is owed, a creditor must be able to call upon the state for help."); TERESA A. SULLIVAN, ET AL., AS WE FORGIVE OUR DEBTORS 3–4 (1989) ("The granting of credit ultimately rests upon the promise of the state to enforce debts. As Henry Clay pointed out almost 150 years ago, the state has always qualified that promise by restricting how far it will coerce people to pay."). *Cf.* Ian R. Macneil, *Contracts: Adjustment of Long-Term Economic Relations Under Classical, Neoclassical and Relational Contract Law*, 72 NW. U. L. REV. 854, 879 (1978) (When planning by parties to a contract fails to focus on maintaining a contractual relationship in the face of conflict, effective contract law remedies may combine with each party's uncertainty of the legal correctness of his position to encourage continuation of the relationship while disputes are resolved.).

Not all observers of collection practices, however, have found creditors' remedies inadequate. Many of them have criticized collection law for the unjustifiable harm it may inflict on debtors who are the subjects of judicial and self-help collection remedies. These observers have articulated a "lost-value" thesis—that certain creditors' remedies impose costs on debtors that do not inure to the benefit of the creditors invoking them.[11] Their critique, a significant one, has motivated re-evaluations of many creditors' remedies that supplement the ordinary judicial collection process. The banning of some of these remedies by federal and various state lawmakers has often been premised on this concept of lost value.[12]

A second aspect of "lost value" in the collection process, which has not been as well explored and articulated by lawmakers and commentators but which is examined in detail in this work, also contributes, I believe significantly, to our discontent with collection law. Lost value in the form of diminished aggregate recovery of all creditors may result from the failure of creditors to switch from coercive to cooperative methods of collection in appropriate cases. Failure of creditors to cooperate with debtors and each other by extending terms of payment in promising workouts may reduce creditors' aggregate recovery, with no offsetting gains, and frequently greater losses, to debtors.

Considering the adverse effects of both aspects of lost value, it is hardly surprising that lawmakers and commentators have focused their efforts on the quest for more humane and efficient practices. What is noteworthy, however, is that these analysts have failed to assess the impact on law reform of certain changes in collection practices by extenders of consumer credit. Because these changes address both aspects of lost value in the collection process, they afford a new perspective for evaluating the efficacy of creditors' remedies—one from which further effort to balance the rights of debtors and creditors should proceed.

---

11. *See* Robert E. Scott, *Rethinking the Regulation of Coercive Creditor Remedies*, 89 COLUM. L. REV. 730, 734–36 (1989) (tracing the history of the lost value concept and the debate it has engendered).

12. A significant example of action by lawmakers premised on lost value is the Federal Trade Commission's promulgation of the Credit Practices Rule. The FTC, relying on cost-benefit analysis, ruled the use of certain contract terms affording creditors additional remedies to be an unfair practice. In the words of the FTC, "[t]he action we take today...is premised on our finding that the cost of each rule proposal [banning contract terms providing for certain remedies] is lower than the costs, to consumers and competition, of the specific practices at which the rule is aimed." FTC Trade Regulation Rule; Credit Practices, 49 Fed. Reg. 7740, 7743–45 (1984) (codified at 16 C.F.R. § 444)(1993).

The changes in collection practices, the analysis of which will occupy center stage in this study, are the result not of governmental edict but of a private-sector response to both aspects of the problem of lost value. The development of this private-sector response has been barely audible over the din of skirmishes for consumer-credit reform on all fronts of government. Action in legislative, judicial and administrative arenas has often been characterized by debates over proscriptions or modifications of various creditors' remedies.[13] The private-sector response to the frequent flash point of collections has been to facilitate voluntary extensions by creditors of their debtors' obligations in instances in which such extensions hold forth the promise of greater recovery of creditors' claims and comport with the desires of the debtor. Hence, the private-sector response remedies defaults, not by greater or lesser degrees of coercion, but by agreements stemming from the same impetus that drives all market transactions—mutual benefit of the parties.

The protagonists in cultivating cooperative effort in the collection of small debts are the various non-profit Consumer Credit Counseling agencies affiliated with the National Foundation for Consumer Credit. These agencies, which exist throughout this country and are now accessible to most Americans, obtain needed extensions in the debtor's payment terms in instances in which both creditors and the debtor are likely to benefit from the extensions.

I focus on the work of the non-profit credit counseling agencies because of the extent of their operations, their dramatic growth rate, the relative uniformity of their operations and the availability of information concerning them. Non-lawyer commercial debt adjusters, who arrange consumer debts for a fee, are also active in those states that do not prohibit the business of debt adjustment. Many states do, however, while others variously supplement the regulatory forces of the market.[14] It seems likely that a non-profit agency that is affiliated with a national trade association will find it easier than a commercial agency to establish the local and national creditor acceptance and support required for successfully operating a debt-adjustment service. The impediments to engaging in commercial debt adjustment, however, do not necessarily preclude some role for the commercial adjuster of consumer debts. Accordingly, the focus of this book on the activities of non-profit organizations should not be construed as a rejection of any role for commercial adjusters of consumer debt.

---

13. *See, e.g.*, MICHAEL M. GREENFIELD, CONSUMER TRANSACTIONS 524–640 (2d ed. 1991); PAUL B. RASOR, CONSUMER FINANCE LAW 489–615 (1985).

14. *See* Carl Felsenfeld, *Consumer Credit Counseling*, 26 BUS. LAW. 925, 927–31 (1971).

16   Games Creditors Play

The first of the non-profit agencies formed with the support of creditors appeared in this country four decades ago. The consumer credit counseling service in Columbus, Ohio, in existence since 1955, claims to be the oldest in service. But the agency in Phoenix contests this assertion. Apparently, credit counseling agencies have learned to assert rival claims to first-in-time priority from the creditors with whom they deal.[15] Profit motivated firms engaged in debt adjustment, however, before the advent of the non-profit agencies.[16] What is significant here is that consumer credit counseling agencies, and especially the presence of these agencies in large numbers over a wide geographic area, are a relatively recent development, spurred on no doubt by the rapid growth of consumer credit following World War II.

By early 1996, Consumer Credit Counseling offices of agencies affiliated with the National Foundation for Consumer Credit were located in over 1200 cities and towns in the United States and Canada, having increased sixfold from only some 200 locations in 12 years. Many of these offices perform credit counseling only, while others are part of multi-service agencies such as Family Services or Catholic Charities. Officials of the National Foundation for Consumer Credit estimate that 90% of the U.S. population now has reasonable access to a main or branch credit counseling office. These officials estimate that during 1996 over 1.8 million debtors will contact member counseling agencies for assistance in avoiding bankruptcy and that those agencies will counsel 972,000 of these debtors. Of those counseled, 34% will be assisted by the implementation of debt management plans in which the agencies obtain creditors' acquiescence in extending payment terms, disburse the debtor's payments to creditors and otherwise administer the workouts they sponsor; another 34% will be instructed by the agencies in self-help through learning budgeting skills; 25% will be advised that some additional measure such as obtaining additional income through another job will be required before a debt management plan will be feasible; and the remaining 7% of debtors counseled will seek relief through bankruptcy. The 972,000 debtors that will be counseled in 1996 are a substantial increase in the 254,000 counseled only eight years earlier in 1988. Moreover, according to the data provided by the National Foundation, the number of debtors seeking the services of credit counseling offices during this eight-year period has grown at a faster rate than that of debtors seek-

---

15. Interviews with various credit counselors who have long-term ties to the counseling movement. I cannot attest with certainty whether Columbus, Phoenix or some other city pioneered the credit counseling movement.

16. *See* Felsenfeld, *supra* note 14, at 926–29.

ing relief in bankruptcy, where 526,000 petitions were filed in 1988 and some 1.1 million are expected to be filed in 1996. Finally, active debt management plans administered by the agencies have grown dramatically in number and dollars returned to creditors in recent years. The 418,000 active debt management plans administered by the agencies in 1996 is a dramatic increase in the 49,739 active plans they administered in 1983, and the dollars returned to creditors from those plans has increased from 242 million in 1987 to an estimated 1.575 billion in 1996.[17]

As the foregoing data reveal, in many instances consumer credit counseling agencies need provide no further assistance to debtors than helping them establish workable budgets. The description that follows, however, focuses on the additional task that data revealed the agencies perform when dealing with more seriously troubled debtors, those for whom budgeting alone is inadequate.

In this role, the agencies obtain creditors' acquiescence in plans that extend the debtor's time for payment. The extensions must be necessary for the debtor's rehabilitation and the debtor's proposed performance under the plan must be feasible. Counseling agencies only attempt extension plans after the debtor and counselor have fully explored the debtor's financial affairs and have determined that payment of all but long-term debt, such as the debtor's home mortgage and sometimes his car note, can be made over a period that customarily does not exceed four years. The agency will sponsor the workout only when the counselor believes that the debtor is committed to the plan and able to live within its confines.

Counseling agencies also administer the plans they sponsor. This duty includes disbursing the debtor's payments to creditors and monitoring the performance of the parties. The debtor must fund the plan as agreed, although obviously counselors will work further with debtors who fall behind in payments so long as they think these debtors are seriously committed to the workout and have the means to fund it. Creditors must not undermine the workout by continuing their individual recovery efforts. The role for these counselors in assisting certain troubled debtors and their creditors is readily suggested by an episode in which one debtor's efforts to go it alone were unsuccessful.

---

17. The data in this paragraph were derived from a November 1996 telephonic interview with and printed materials provided me by Durant S. Abernethy, president and CEO of the National Foundation for Consumer Credit. A further interview by phone in December 1996 with Bill Furmanski, Director of Communications of the National Foundation for Consumer Credit, and Melony Branbon, assistant to Mr. Abernethy, helped me interpret some aspects of the material I had been furnished. Any errors in the presentation of the data are, of course, my own.

B. Creditors' Self-Inflicted Wounds in Microcosm—Kelly's Case

Even the most avid reader of advance sheets or on-line computer services providing reports of cases decided by the courts will be unfamiliar with the report of Kelly's case that follows because none of the parties in that collection tragedy ever invoked legal process. I justify my effrontery in usurping the roles of two major institutions—a duly constituted court and the West Publishing Company, which supplies lawyers with court opinions—on the grounds that collection tragedies, like Kelly's, often occur in a non-judicial setting. Those that do find their way into cases reported by courts seldom analyze in any detail the ultimate issues with which I am concerned here. Relating Kelly's sad story in some detail provides a necessary background for the subsequent analysis of creditors' behavior with which this work is concerned.

Although my report of Kelly's case is derived solely from interviews with Kelly, and any student of controversy knows the dangers of hearing only one side of an issue, I quizzed her at length and believe my report is substantially accurate. But if Kelly's case is misstated in some respects, what is important is that there are cases, probably numerous ones, that do reflect the facts that are material to the analysis that follows.[18]

After more than three months of unemployment, Kelly's relief in finding a job was short-lived. While her earnings were modest, the demands of her creditors for full payment on all accounts in arrears were not. The threats of a secured lender to repossess her car and the electric company to discontinue service easily leveraged their claims into priority. Meanwhile, she faced the growing discontent of two banks in a city 300 miles away, which had issued her credit cards. Their collection departments telephoned frequently and tried to exact large checks, even if post dated. In time, insinuations of vaguely unpleasant consequences that would accompany continued arrearages gave way to veiled and then direct threats to garnish wages and send damaging reports to credit bureaus. Some of the actions of Kelly's creditors may have violated statutory and common law norms,[19]

---

18. To protect Kelly's right of privacy, I have changed her given name, omitted her surname and changed certain identifying details in my report of her case.

19. The Fair Debt Collection Practices Act (Title VIII Consumer Credit Protection Act, §§ 801–818, 15 U.S.C. §§ 1692–1692o (1988), regulates informal or nonjudicial collection practices. But this act applies only to actions of third-party debt collectors. Some states, however, have extended the scope of their various legislative regulations of the debt collection process to include the actions of a creditor collecting its own claims. *See*

but the usual impediments to bringing judicial action loom even larger to an impoverished debtor than to other would-be plaintiffs. Besides, Kelly wished to pay her debts, not sue her creditors.

By turns embarrassed by her plight and resolute in her catch-up efforts, Kelly worked at convincing her creditors that her need and perseverance warranted their acquiescence in extended terms for payment. Creditors received small payments with promises of larger ones as her condition improved. Things were far from hopeless. Barring further interruption of income or non-volitional expenses, both unforseen events, Kelly could have paid her obligations under the plan she had worked out in some 18 to 24 months.

Her plan did necessitate a no-frills budget. Loss of a contact lens meant resurrection of an old pair of glasses or suffering impaired vision for the one occasion she felt glasses inappropriate—her wedding. Despite the quality of her performance and the extent of her explanations, the banks adamantly refused to extend her terms for payment and relentlessly demanded more than she was able to pay. Shortly after Kelly and her equally impecunious husband had their checking account closed due to overdrafts, they moved without providing her creditors a forwarding address. In the parlance of collectors, Kelly "skipped."

While she may find new employment and resume payments, both her willingness and ability to do so are not likely to have been enhanced by her experience. A creditor may think it worthwhile to "skip trace" Kelly and sue if she can be found. But any amount netted after recovery of expenses may be less than that which would have been voluntarily forthcoming if her creditors had acquiesced in her workout plan. A more serious flaw in the pursuing creditor's calculus could result in the additional costs of that creditor's coercive collection effort exceeding any recovery from Kelly.

A creditor's attempt to collect by legal process from the debtor may be thwarted by exemption laws[20] and by the debtor's secretion of non-exempt assets until cost-conscious creditors are discouraged from further effort. Creditors and their collection agencies may use self-help or judicial remedies such as discovery depositions, creditors' bills and proceedings supplementary to execution to discover debtors' assets. No discovery proce-

---

ROBERT J. HOBBS, NATIONAL CONSUMER LAW CENTER; FAIR DEBT COLLECTION 423–28 (1987) (chart of state debt collection statutes). Moreover, tort remedies may also provide redress to injured consumers. *See id.* at 215–28.

20. *See generally* STEFAN A. RIESENFELD, CASES AND MATERIALS ON CREDITORS' REMEDIES AND DEBTORS' PROTECTION 289–333 (4th ed. 1987) (exploring some of the more common provisions of exemption laws that vary greatly among the various jurisdictions of the United States).

dures are costless, however, and creditors are understandably reluctant to risk throwing good money after bad.[21]

Should Kelly feel trapped again, she may embark on yet another journey to yet another community. At this juncture, however, she will more likely go to a Federal courthouse. There she will seek and probably obtain discharge of her debts in Chapter 7 of the Bankruptcy Code,[22] eschewing, because of her earlier tribulations, bankruptcy's provisions for a court protected partial or full workout in Chapter 13.

Because Kelly has no significant assets, the most obvious cost of her discharge in Chapter 7, the forfeiture of her non-exempt property,[23] will be minimal. Kelly's paucity of resources also furthers, however, her ability to play hide-and-seek with her creditors and escape their efforts at collection without needing to invoke the protection of the bankruptcy court.

All this conjecture, which probably seems too obvious to those familiar with collections to warrant exposition, does have a purpose—to reveal a paradox worthy of serious exploration by students of debtor-creditor law. If Kelly's failure to work her way out of debt was attributable in no small measure to the inflexible stance of her creditors, and if those creditors will likely suffer loss as a result, why did these supposedly rational actors behave in such a seemingly counterproductive way?

Where the question is addressed to a discrete case, as it is here, an easy answer is suggested. Kelly's creditors may have been privy to information that, based on their experience in collection matters, caused them to discount the present value of her promised performance to an amount less than their anticipated recovery from continuing their coercive collection efforts.

But Kelly's case is not reported for the humbling but mundane purpose of suggesting that the collectors may have been better informed or simply exercised better judgment than the case's reporter on these pages. I concede that those collectors probably had more experience in dealing with troubled debtors than I. Doubtlessly, in many cases creditors obtain increased individual and aggregate creditor recovery by applying unrelenting pressure on debtors who need to turn their financial affairs around before it is

---

21. *See* JAMES W. MOORE & WALTER R. PHILLIPS, DEBTORS' AND CREDITORS' RIGHTS 2-169–2-216 (4th ed. 1975).

22. The grounds for denial of discharge in Chapter 7 bankruptcy are limited primarily to cases in which the debtor has been guilty of certain misconduct or has previously obtained a discharge in a Chapter 7 case commenced within six years of filing the petition initiating the subsequent Chapter 7 proceeding. *See* 11 U.S.C. § 727(a) (1993). Certain debts may be excepted from the Chapter 7 discharge. *See* 11 U.S.C.A. § 523(a) (West Supp. 1996).

23. *See* 11 U.S.C.A. § 522 (West Supp. 1996).

too late. Moreover, there are debtors who are going to fail, and by their efforts, some creditors will get more than others. But there are other debtors against whom such pressure is counterproductive, and Kelly is offered as typifying the latter.

Nor do I want to explain Kelly's case as an anomaly—that each gunslinger misconstrued the facts and fired too soon, shooting himself in the foot. I think the injury to Kelly's creditors could have occurred absent aberrant creditor conduct. While Kelly's case faithfully relates—even down to the melodramatic, so suspect "contact-less" wedding—some miserable events in the life of but one individual, her case is meant to be generalized. I believe there are numerous instances in which creditors press debtors beyond the breaking point when it would appear that their aims would be better served by entering into cooperative ventures allowing extended payments.

Of course, not all cases in which extensions seem justifiable, but are not given, result in failure of debtors to repay. The author of a guest opinion column in *Newsweek* relates a story similar to Kelly's but with a happier ending. Elizabeth Hudson lost her job as a press aide, and, in a state of panic, initially attempted to stonewall her creditors by ignoring their requests for payment. She reported that "[s]ome collectors sounded as pleasant as a salted sea slug." After sharing the secret of her job loss and dire financial straits with one sympathetic creditor, she tried explaining her situation to other creditors and promising them full payment later. Although some cooperated, many continued to press for immediate payment. Elizabeth Hudson survived financially, later reentered the work force and subsequently paid her bills.

While the coercive collection practices of some of Hudson's creditors did not result in loss on aggregate creditors' claims in this instance, their coercive practices may still have resulted in unnecessary costs. These costs extend beyond the most obvious, Hudson's unnecessary suffering. Ms. Hudson reports that except in emergencies, she now pays cash. Thus, in some small measure, even in this better than worse-case scenario, the credit industry still suffers.[24]

Obviously, no claim to originality accompanies my observation that creditors' coercive collection practices may produce less recovery than creditors' cooperation in extending payment terms. Some two decades ago, the late Yale law teacher Arthur Leff characterized "American collection mechanisms and institutions" as "grossly inefficient engendering huge amounts of unnecessary grief and loss for all participants,"[25] and certainly, even then, Leff's indictment was scarcely novel.

---

24. Elizabeth Hudson, *I Don't Need All the Credit,* Newsweek, July 14, 1986, at 9.

25. Arthur A. Leff, *Injury, Ignorance and Spite—The Dynamics of Coercive Collection,* 80 YALE L.J. 1 (1970).

Recast in general terms, the question posed by Kelly's case becomes why do creditors seemingly act irrationally in a significant number of collection cases? So generalized, the question suggests its answer. When commonly observed market practices fail to comport with theory, the theory is embarrassed, not the actors in the market. The theory, thus far, lacks a more sophisticated explanation of the dynamics of the collection process, one that will explain what are otherwise examples of seemingly irrational creditor behavior.[26]

C. Kelly's Case Generalized: A Preview of the Causes and Cures of Counterproductive Collection Practices with Ramifications for the Reform of Creditors' Remedies

Chapter Three of this book suggests that the coercive methods often employed by consumers' creditors, which may be destructive of their collective and even individual recovery, are to a great extent the product of a contest among them. To be sure, a concept so long revered by the law as that of rewarding creditor diligence is not without strong underpinnings. Therefore I explore both the individual and collective benefits of a system that rewards the victor, rather than apportions the spoils.[27]

But justifying individual reward in cases in which it increases or does not decrease aggregate recovery fails to explain why the race of diligence among creditors continues when it militates against collective and even individual

---

26. Assuming that Kelly's noncooperative creditors nevertheless recognized that payment extensions by all her creditors would probably increase the amount of their recovery is a useful device for separating one fundamentally important reason for overly coercive collection from others. *See* Chapter 4, Section A (a game-theoretic analysis of creditors' failure to cooperate in this context). But as creditors' failure to cooperate in a workout that would increase their recovery may also result from their failure to perceive this fact, the assumption is removed in other analyses of Kelly's case. *See* Chapter 3, Section C (2) (factors that distort creditors' perceptions) and Chapter 4, Section B (2) (a game-theoretic analysis of the effect of these distorting factors on creditors' collection practices).

27. *See generally*, Lawrence Berger, *An Analysis of the Doctrine That "First in Time is First in Right,"* 64 NEB. L. REV. 349 (1985) (exploring the ancient notion of temporal priority in a variety of legal contexts and concluding that the principle of encouraging economic productivity justifies the rule).

recovery. Thus, Chapter Three also explores why counterproductive coercive practices continue, despite the existence of alternative collection procedures. To do so, that chapter examines both legal checks and private-ordering alternatives to the rigors of unbridled competition among creditors. Chapter Three then provides some explanations of why traditional means of shifting from competitive to cooperative recovery, often successfully employed in cases of commercial debtors, have been often unsuccessful in consumer settings.

My emphasis in Chapter Three on the harmful aspects of creditor competition is not meant to impugn the theory that asymmetries of information between the individual creditor and his debtor are a significant factor in explaining destructive collection practices. From its beginnings in the work of the late Arthur Leff of Yale,[28] through the contributions of Wisconsin's William Whitford[29] and Virginia's law dean Robert Scott,[30] the "lost-value" thesis has attributed debtor loss that does not result in creditor gain to the failure of creditor and debtor to possess common, accurate information concerning their viable alternatives.

Traders extending credit throughout most of humankind's long existence have had much easier access to that information. Chapter One's prehistoric traders of spear points and bear skins lived together in a small clan and in any credit transaction between the two knew what the other was doing in comparison with what he was capable of doing to remedy any default. Our knowledge of each other's efforts and capabilities is not nearly so easy to come by in today's mass markets. Certainly creditors can and do gather facts and assess them in collection cases, but information is not costless. Operating under the cost constraints imposed by the smaller claims that normally distinguish consumer from commercial cases, creditors often fail to distinguish instances in which extensions hold forth the promise of greater recovery from those in which extensions will probably be counterproductive.

The failure of Kelly's creditors to acquiesce in her attempted workout may well have been influenced by inadequate or misleading information on each of their parts concerning her need for concessions and her desire and ability to turn her financial affairs around. But this asymmetry of information between creditor and debtor does not complete the explanation of the role of informational deficits in financial debacles such as Kelly's. It more likely pulled in tandem with another.

---

28. Leff, *supra* note 25.
29. William C. Whitford, *The Appropriate Role of Security Interests in Consumer Transactions*, 7 CARDOZO L. REV. 959 (1986); Whitford, *A Critique of the Consumer Credit Collection System*, 1979 WIS. L. REV. 1047.
30. Scott, *supra* note 11.

While an individual creditor's belief in the debtor's needs, desires and abilities are necessary conditions to obtaining concessions from that creditor, they are not always sufficient ones. In most cases of serious overextension, the creditor must also know that the other creditors of the debtor harbor similar beliefs, and that they too will cooperate in a workout. Furthermore, because no workout is without risk of failure, each creditor will want the other creditors to cooperate on some basis commensurate with her own allowances to equitably share that risk of failure. Finally, the creditor must have some cost-effective means of monitoring for problems that may arise during the workout—not only for the effect that breakdowns may have on the debtor but also for their effect on the actions of competing creditors. Overcoming these impediments to securing necessary cooperation among creditors imposes significant additional transaction costs that, along with those attributable to assessing the debtor's need, commitment and ability to effect a workout, must be contained, lest they defeat efforts at otherwise promising workouts.

Information deficits and asymmetries are indeed at the heart of harmful creditor competition, but, except in a one-creditor world, they extend beyond each creditor's uncertainties about the debtor's actions. Any attempt to explain overly coercive collection must also reckon with an individual creditor's rational fears that other creditors will engage in strategic behavior, coercing the debtor to pay their claims in preference to that of the creditor granting the concession and scoring their recoveries at that creditor's expense.

Competition among creditors is therefore central to understanding coercive practices that may diminish aggregate recovery. Yet, this phenomenon of creditor competition has received scant attention in previous analyses of consumer-credit collection practices. In Chapter Four, I will explore the phenomenon in some detail by employing elementary game-theoretic decision-making analysis. Two principal concepts of game theory assist in understanding the role of creditor competition in harmful collection practices.

When the benefits that will result from all creditors abandoning the race of diligence are clear, the obstacles to overcoming creditor competition bear a striking parallel to the barriers to cooperation encountered by the participants in the classic game of Prisoner's Dilemma. Aggregate creditor recovery is highest when all creditors cooperate in the workout by extending payment terms. But a creditor who cooperates while another employs a coercive collection strategy will recover the least amount of her claim, even less than she would have obtained if both creditors had used coercive strategies. The creditor who uses coercion against the others cooperation scores the highest recovery of all, even more than she would have received if both creditors had cooperated, but she does so at a significant cost to aggregate creditor recovery.

Before the impetus of higher total creditor recovery dictates mutual creditor cooperation, each creditor must have some means of assuring herself that other creditors will cooperate in the workout. In the absence of assurance of mutual cooperation, game theory teaches that each creditor's best strategy is coercive, not cooperative. Creditors can and do get together in commercial cases where large claims are at stake, but in many consumer cases, transactions costs act like stone walls and prison bars in the prototype of the game of Prisoner's Dilemma and keep the creditors from reaching an agreement to cooperate in a workout.

An additional hurdle exists when creditors erroneously conclude that the debtor will not use concerted creditor extensions if forthcoming to increase their aggregate recovery, when they believe that the only question is how distribution from a finite pool of assets devoted to payment of claims will be made among them. There, the lesson of zero-sum games is that each creditor will try to outdo the other in employing coercive collection measures. Engaging the tools of game theory in Chapter Four, even in only an elementary fashion, reinforces and elaborates on the conclusions reached independently by Chapter Three's exploration of the role of the creditors' race of diligence in thwarting beneficial cooperation among creditors.

Game theory is also useful in obtaining valuable insights on cost-effective solutions to the creditors' problem. Chapter Five explores methods suggested by studies in applied game theory. In the context of these studies, I examine the role that an agent in the private sector supported primarily by creditors—the many nonprofit Consumer Credit Counseling Service organizations affiliated with the National Foundation for Consumer Credit—plays in overcoming the formidable obstacles to cooperation examined in the preceding chapter. Economies of scale in instituting and monitoring promising workouts assume an instrumental role in explaining how counselor-assisted debt adjustment functions within the severe cost constraints that are indigenous to consumer credit. Chapter Five also addresses how the work of Consumer Credit Counseling Services differs from that of a public agency designed to secure the cooperation of creditors in debtors' rehabilitation efforts—the bankruptcy courts administering Chapter 13 plans.

Chapter 13 bankruptcy and counselor-assisted workouts may be used interchangeably, but in practice the two customarily function far differently. Counselor-assisted workouts propose full payment to all creditors, while average proposed payments to unsecured creditors in Chapter 13 proceedings are considerably less. Moreover, debtors in counselor assisted workouts are better motivated to complete their plans, for unlike Chapter 13 proceedings, no discharge is obtained unless they do.

What meaning does the mounting success of this private-sector initiative, which, unlike its commercial credit counterpart, eschews any significant role for lawyers, hold for those of us whose concern is law? There are important lessons for law reform inherent in the now palpable presence of an agency that enables the parties to both curb the debtor's lost value and maximize creditors' recovery in cases that come within the agency's purview.

Chapter Six explores some ramifications of the availability of debt adjustment through consumer credit counseling services for most Americans. Because qualifying debtors may now resolve their financial problems through counselor-assisted workouts and are no longer forced to choose between the sting of coercive collection or the stigma of an unwanted bankruptcy, the case for restricting creditors' remedies to prevent their harmful use against debtors who are fruitfully pursuing workouts is considerably diminished.

Overextended debtors now have the power to block counterproductive collection methods in virtually all instances. For debtors who can pay their obligations in full with extensions in time for payment, there is consumer credit counseling. For those who need more than extensions, the law provides two forms of bankruptcy relief: Chapter 13 for those who are willing and able to devote some future effort to payments to their creditors and Chapter 7 for those who are not. For debtors, however, who are able with adjustments in spending habits to pay their obligations but who fail to do so, creditors need effective remedies. Chapter Six of this work concludes therefore that the trend of recent years should be reversed: that creditors need more rather than less effective remedies as the principal use of these remedies will be against debtors who ought to be held accountable for their obligations.

My conclusion that creditors' remedies should be reinforced is also based on the work of commentators who find the lost-value premise often wanting as a justification for banning creditors' remedies. Robert Scott's attack on the premise need not deny, as do those of some other commentators, that debtors' losses at the time their property is taken may exceed creditors' gains. To the contrary, according to Scott the potential for lost value produces a benefit by providing a means of signaling debtor commitment that compensates for deficiencies in legal enforcement of claims. The lost value that the debtor may suffer from the creditor's use of the remedy commits the debtor to pay the obligation. By focusing on the time when the credit transaction is contemplated, Scott finds that the potential for lost value makes mutually advantageous credit transactions, which might not otherwise transpire, doable.[31] Nevertheless, debtor loss that does not result in equivalent creditor gain at the time the debtor's property

---

31. *See id.* at 734–56.

is taken has currency among lawmakers as a justification for banning some longstanding creditors' remedies.

The remedies that have been attacked by lawmakers are ones that supplement the ordinary judicial collection process. These attacks frequently restrict the taking of security interests—consensual liens that accord the creditor certain rights in the property subject to the lien—or they ban certain of the remedies that otherwise arise from the taking of security interests. These supplemental remedies, not all of which are associated with security interests, are valued by creditors because of the inadequacies that they perceive in the ordinary judicial collection process that may be used in the collection of any claim.

Exemption laws, the effects of which creditors may avoid by taking security interests in property otherwise immune from creditor process, often unduly restrict the effectiveness of the judicial collection process under which an unsecured creditor must proceed. Laws in some states that impose no limitation on the value of exemptions in certain types of property are a mockery of our law of civil obligation, permitting a debtor to shelter literally millions of dollars of assets from the claims of creditors. Chapter Six concludes therefore that where such exemption laws exist, creditors have even more pressing need for the supplemental remedies that have been under siege. But not all transactions lend themselves to the taking of security interests, or at least not to ones that fully protect the creditor, and so an additional need exists to address exemption laws that go far beyond their commonly recognized purpose of insuring that the debtor does not lose the basic necessities of life.

3

Collection Practices: Competition v. Cooperation Among Creditors

A. Coercion as the Product of a Race of Diligence Among Creditors

(1) The Competitive Nature of Collection

While collection methods vary with the needs of the case and the sophistication of the collector, all cases in which the debtor is seriously in arrears share one characteristic: creditors' actions are designed to coerce payment from debtors who have not responded to mere requests to honor their obligations. Since the abolishment of body execution or imprisonment for market debt, legitimate coercive collection has consisted of either seizure of debtors' property under legal or self-help process or creditors' threats to use those processes to exact payment from their debtors. Coercive remedies are needed to show that the legal obligation upon which the institution of credit rests is not illusory, that the power of the state may be brought to bear, at least to some extent, to enforce claims against recalcitrant debtors. But understanding why coercion figures so prominently in the collection of consumer debt also requires recognition of an additional force that drives the collection process—competition among creditors.

Collection cases generally cast creditors in the role of adversaries, for creditors of an overextended debtor compete with each other for their debtor's limited resources. When a debtor's circumstances preclude payment of all obligations, some creditors by their efforts may get more than others. Competition results in each creditor attempting to convince the debtor that failure to honor his claim will leave her worse off than failure to honor the claim of any of her other creditors. Moreover, as this work will show, competition among creditors functions not only as a significant factor in generating coercive collection but also as the principal obstacle to eliminating it in those cases where coercive measures are counterproductive to the recovery of creditors' claims.

Have I overstated the role of competition among creditors in producing coercive collection? Obviously creditors do not compete with each other in all collection cases. Where a debtor has the resources to pay an undisputed obligation yet refuses to do so, the holder of that obligation is not concerned with the ploys of fellow creditors but with those of the debtor. The legislative history of the federal Fair Debt Collection Practices Act indicates, however, that a debtor does not ordinarily attempt to avoid payment of undisputed debts that the debtor can pay. That legislative history reveals that there is "universal agreement among scholars, law enforcement officials, and even debt collectors that the number of persons who willfully refuse to pay just debts is minuscule."[1] Thus debtors fail to pay their obligations because they have inadequate resources, and creditors, some of whom will fail to recover their claims, compete with each other by the use of coercive collection measures for those resources that the debtor does possess.

While this statement is no doubt true, it does not provide a full explanation of the principal cause of debtor default and the associated reason for creditors' use of coercion. Observing that debtors have inadequate resources for which creditors necessarily compete fails to address the fact that creditors' coercive collection efforts often do more than determine distribution among creditors from a finite pool of resources. Creditors' collection efforts often increase the amount of resources that the debtor commits to payment of creditors' claims. Some debtors, at their creditors'

---

1. S. Rep. No. 382, 95th Cong., 2d Sess. 2 (1977). The complex issue of whether a significant number of even those debtors who have filed for bankruptcy are "won't-pays" as opposed to "can't pays" has sparked a lively debate. *Compare* Credit Research Center, Krannert Graduate School of Management, Purdue University, Monograph No. 23, Consumer Bankruptcy Study Vol. I, Consumers' Right to Bankruptcy: Origins and Effects 88–90 (1982) [hereinafter Purdue Study] *and* A. Charlene Sullivan, *Reply: Limiting Access to Bankruptcy Discharge*, 1984 Wis. L. Rev. 1069 *with* Teresa A. Sullivan et al., As We Forgive Our Debtors 199–29 (1989) [hereinafter Sullivan Et Al, As We Forgive Our Debtors]; Teresa A. Sullivan, et al., *Rejoinder: Limiting Access to Bankruptcy Discharge*, 1984 Wis. L. Rev. 1087 (1984) *and* Teresa A. Sullivan, et al., *Limiting Access to Bankruptcy Discharge: An Analysis of the Creditors' Data*, 1983 Wis. L. Rev. 1091 (1983). The line between inability and unwillingness to pay is blurred, however, to the extent that inability to pay is a function of failure to curtail discretionary expenses or to fully exploit earning potential or other resources. Although a debtor may not have the ability to pay, creditors must justify their collection efforts by assuming that the debtor may be able to provide some payment to one or more of the creditors. Therefore, it seems that the element of creditor competition virtually always exists in collection cases.

urgings if not by their creditors' direct use of legal process, will strive mightily to obtain additional resources and to free others for debt payment that were previously committed to the debtor's current consumption. Therefore, recognition of the role of creditor competition in producing coercive collection practices does not rule out an additional cause of these coercive practices—that of modifying debtor behavior in order to increase the resources committed to payment of creditors' claims.

Creditors tell me that they would attempt to alter the debtor's income and expense patterns whether they believed they were the debtor's sole creditor or one of many. I do not question the truth of this statement, but I also have no doubt that creditors realize that if their coercive efforts fail to turn the debtor's financial affairs around so that all creditors are paid—and there are many cases in which they fail to do so—the creditors who will take the least losses are those who have scored the greatest recoveries before their debtor's efforts failed. And as greater recovery is commonly associated with the strength of a creditor's demands relative to those of other creditors, we return to the role of creditor competition in imparting a significant if not principal thrust to each creditor's use of coercive collection.

(2) Establishing Creditor Priority by Applying Formal Rules of Law

Collection law in its formal setting of judicial proceedings has long recognized the role of competition among creditors and, indeed, has fostered that competition by the rule it most frequently employs in determining priority among competing creditors' claims to assets of the debtor. That rule, "first in time is first in right," promotes what has long been recognized as "a race of diligence among creditors."[2]

The temporal event upon which a creditor's priority is based contributes significantly to furthering the competitive aspects of collection. Priority is based, not on the time of acquisition of a claim against the debtor, but on the time of acquisition of a lien or other interest that is good against subsequent claimants to the property in contest.[3] Thus the rule of priority,

---

2. *See, e.g.*, THOMAS H. JACKSON, THE LOGIC AND LIMITS OF BANKRUPTCY LAW 8–9 (1986); ROBERT L. JORDAN & WILLIAM D. WARREN, BANKRUPTCY 392 (3d ed. 1993).

3. *See e.g.*, Jackson, *supra* note 2; Stephan A. Riesenfeld, *Collection of Money Judgments in American Law—A Historical Inventory and a Prospectus*, 42 IOWA L. REV. 155, 159–60. During the Great Depression, Dean Sturges of the Yale Law School felt, however, that there were advantages to according priority on the basis of acquisition of claims, rather than acquisition of liens, when the competing creditors were unsecured at the time legal proceedings were initiated. *See, e.g.*, Wesley A. Sturges, *A Proposed State Collection Act*, 43 YALE L. J. 1055 (1934). He perceived a need to require "lenders and sellers on

which awards an earlier lien holder first claim on the property to which his lien attaches, creates a world in which creditor status is fluid. A mere general or unsecured creditor, one without a lien on any of the debtor's property, can obtain upward economic mobility by acquiring secured status through judicial action producing a judgment, execution, garnishment or attachment lien. Moreover, the creditor is motivated to act quickly, before all the debtor's property is encumbered by the liens of other creditors. A creditor with this motivation is certainly disinclined to grant the debtor extensions for payment.

Of course a transfer of cash by the debtor to the creditor—the happy event of payment of the creditor's claim—is to be preferred to a lien upon even the most marketable of property, for payment permits the creditor to retire unscathed from the field of battle. Here too, however, the first-in-time rule applies. Except in cases of transfers in fraud of creditors, payments to employees, cash sellers, other creditors, and any other good-faith purchaser for value will remove the funds with which those payments are made from the grasp of a creditor previously bent on collecting her claim from them. Expeditiousness in acquiring payment or lien status then is the factor that determines how individual creditors fare when their debtor fails.

Creditors who plan their credit extensions initially around consensual liens or creditors who are the beneficiaries of transactions giving rise to liens arising by operation of law acquire their secured status, and thus establish the date from which their priority is based, earlier than creditors who must await the judicial collection process to acquire a lien upon some property of their debtor. Liens produced by judicial action often arise only shortly before execution sale marks the end of the collection process.

---

credit to recognize a duty to the existing creditors of a prospective borrower or buyer on credit not to burden the debtor beyond his ability to pay," *id.*, and proposed model legislation that would accomplish that end, with certain exceptions, *id.* at 1063–64, by a system allowing creditors to "underfile" not only in traditional collective actions such as bankruptcy but also in proceedings initiated by an individual creditor solely for the purpose of collecting her own claim, *id.* at 1058–60. Sturges' proposal has not been successful in replacing the general rule that establishes priority on the basis of lien acquisition, perhaps because he was unsuccessful in refuting the argument that his proposal would disadvantage debtors too severely in their attempts to acquire further credit. *See id.* at 1056–57. It is noteworthy, however, that he traced harmful aspects of creditor competition back to a time earlier than the onslaught of collection contests. Although Sturges wrote during the Great Depression, when the number of lenders to consumers and small business firms was significantly less than it is today, he recognized a deleterious aspect of creditor competition in credit-granting practices.

Timely taken and recorded purchase-money consensual liens, such as a mortgage on the debtor's home or a security interest in the debtor's automobile, establish the creditor's claim on the collateral from the moment credit is extended. The automobile repair shop may also enjoy lien status from the time its claim arises. Repairers of goods have long been given liens on the property on which they work, and these liens arise merely from the status of the parties to the transaction. They require no manifestation of intent on the part of the debtor to create them. The justification for such liens, which arise by command of statute or common law and hence are known as liens arising by operation of law, is that a simultaneous exchange of consideration such as that which may occur in a sale of goods is not feasible in a transaction in which the performance of one party cannot be rendered instantly. To protect the party performing the repairs, the law provides a lien that is efficiently created and enforced because it attaches to property already in the lien holder's possession.

Because a creditor without a consensual lien or lien arising by operation of law cannot establish her priority in any of the debtor's property until she acquires a lien in the judicial action brought to enforce her claim, a considerable time may elapse from her realization that the collection will be troublesome and the establishment of any lien in her favor. In most cases no lien may be acquired by judicial action prior to judgment being rendered against the debtor. While law provides for prejudgment seizure of property and the prejudgment acquisition of a lien on that property in cases of extraordinary creditor need, the grounds affording unsecured creditors prejudgment judicial remedies of this nature are narrowly drawn. They are limited in many states to cases in which the debtor is removing his property from the jurisdiction or is secreting or fraudulently transferring it so as to defeat efforts to collect any judgment that a creditor may subsequently obtain. Liens acquired by judicial action upon the debtor's property must usually await recording of final judgment or levy on the debtor's property under a writ of execution, which may be issued only after final judgment is entered.

The law's focus on establishing priority by rewarding diligence in the contest among creditors continues until the parties invoke some collective measure such as bankruptcy or an agreement among all creditors reducing the amount that must be paid on their claims or extending the time for payment of those claims, whether reduced in amount or not. Even in cases that end in collective action, however, a creditor may often retain liens and payments that she acquired before the commencement of the collective action signaled the end of the race of diligence. No wonder creditors strive to obtain payment or, failing that, a lien on their troubled debtor's property. Elitism runs rampant in creditor circles, for no one wishes to suffer the fate of the unsecured creditor who normally receives little or nothing in a consumer's bankruptcy proceeding.[4]

Early lien acquisition is obviously also advantageous when the debtor does not seek relief in bankruptcy or enter into some other proceeding that deals with claims collectively. Creditors who belatedly enter the race for lien acquisition often finish out of the money because there is simply no unencumbered property remaining out of which their claims can be satisfied.

A creditor who has acquired a lien by consent of the debtor or by operation of law may enjoy something more, both in and out of bankruptcy proceedings, than the establishment of priority against other creditors sooner than he could have done by legal action resulting in a judicial lien. As exemption laws normally only protect covered property from being subjected to judicial liens and forced sale under judicial proceedings brought by unsecured creditors, the creditor with a consensual lien or one arising by operation of law may have access to property that would otherwise be immune from the reach of creditors.[5]

(3) Establishing Creditor Priority by Informal or Self-Help Collection Practices

The race of diligence among creditors is no less pronounced when focus is shifted from judicial remedies, those controlled by process issued by courts and executed by officials of the state, to informal or creditor self-help collection methods.[6] Some creditors have extraordinary self-help rights that produce the same result as judicial action: sale of property of the debtor to satisfy or reduce the creditor's claim. The most notable of these extraordinary self-help remedies are the right of a secured party with a

---

4. *See generally*, DAVID G. EPSTEIN, DEBTOR- CREDITOR LAW IN A NUTSHELL 256–94 (4th ed. 1991) (summary of the different treatment accorded secured and unsecured creditors in bankruptcy).

5. *See* Jordan & Warren, *supra* note 2 at 62–3.

6. While the authors of some primers on non-judicial collection practices note the problem of creditor competition, *see, e.g.*, DONALD L. HENRY, HENRY ON CREDIT AND COLLECTIONS 129 (1977) (Collecting "is a competitive act with other creditors"); NORMAN KING, PAST DUE: HOW TO COLLECT MONEY 159 (1983) (a collector must know his competition to know why another creditor got ahead of him), the only attention typically given by these writers to the competitive aspects of collection is that implicit in their discussions of collective actions and undertakings. Moreover, these writers typically confine their discussions of general receiverships, general assignments for the benefit of creditors and composition and extension agreements among creditors to cases of business debtors.

consensual lien to use self-help in taking[7] and selling[8] personal property collateral following the debtor's default, and the right of a repairer of goods to retain and sell those goods under a lien arising by operation of law when the customer fails to pay.[9] Ordinarily, however, creditors' self-help remedies do not encompass the right of creditors to help themselves to any of the debtor's property. But this limitation on self-help efforts is of little moment to the seasoned collector.

What Kelly's creditors did was talk, not take. And while some of the talking was about taking, and thus about employing legal process, there is a world of difference in collection methods that employ a deputy sheriff to remove property from the debtor's possession and those that project only an image of that process to strengthen demands for payment. Theorists seldom recognize the priority-establishing aspect of informal or non-judicial collection. One who has, Thomas Jackson, compares the collection process outside of collective proceedings to people buying tickets to a popular rock concert. The best seats go to those first in line, while those at the end of the line may get nothing. Jackson notes a variety of methods that creditors can use to stake a place in the collection line, including not just lien acquisition by operation of law and judicial action but also priority established by " 'voluntary' actions of the debtor: the debtor can simply pay a creditor off or give the creditor a security interest in certain assets".[10] The debtor will often be motivated to take these "voluntary actions" by the strength of the favored creditor's demands.

Certainly, talking contests, like those that occurred in Kelly's case, can and do escalate into cases in which such trappings of legal process as judgment liens on realty, garnishment liens on wages and checking accounts, and execution liens on automobiles and other consumer durables are crucial. But so very many do not, and for good reason. Attorneys' fees and court costs usually loom so much larger relative to the amount of the claim in a consumer case than a commercial one that a much higher probability of recovery is needed to justify bringing the action.[11] Commonly, however,

---

7. *See* U.C.C. § 9–503 (1989) ("In taking possession, a secured party may proceed without judicial process if this can be done without breach of the peace or may proceed by action.")

8. *See id.* § 9–504(3) (judicial foreclosure not required under statutory authority to dispose of collateral "by public or private proceedings").

9. *See* Jordan & Warren, *supra* note 2, at 2.

10. Jackson, *supra* note 2, at 9.

11. *See generally* PAUL ROCK, MAKING PEOPLE PAY 51–75 (1973) (anecdotal evidence of the economic constraints imposed on the English debt collection process). In one self-help book addressed to the small-business operator, the author notes that lawsuits may be costly, long, involved and tricky. He counsels his readers to consider such alternatives as

the probability is not nearly as great. Overextended consumers generally own few unencumbered assets of real market value,[12] and exemption laws, which do not limit collection from corporate debtors in commercial cases,[13] often insulate any property of the consumer debtor upon which levy would be cost-effective.

Reinforcing this disincentive of a creditor to sue, which is based on the creditor's belief that no judicial remedy will be cost-effective in a given case, is the creditor's belief that many debtors hold a different view of the efficacy of judicial remedies.[14] Creditors know that debtors are likely to be better informed of the broad reach of a judicial remedy than a specific limitation on its use. A judicial remedy may therefore be more effective when holstered than when drawn, fired and found to contain only blanks.

In a similar manner, a creditor without some self-help remedy may benefit indirectly from a law that gives that remedy to some other creditor. While many buyers of goods on credit, like Kelly, know that their unattended car may disappear from a street or driveway if they fail to make installment payments on the loan it secures,[15] many buyers do not realize that unsecured credit sellers and lenders, those without the consensual lien that the automobile financer invariably has, have no similar self-help right to recover the goods they have sold or financed.[16] Unsecured creditors may further the

---

using a collection agency or having an attorney write the debtor. *See* MILTON PIERCE, HOW TO COLLECT YOUR OVERDUE BILLS 139–42 (1980).

12. Empirical evidence shows that the mean assets of debtors who have filed for relief in bankruptcy are worth only 30% of the mean assets of the national population. SULLIVAN, ET AL., AS WE FORGIVE OUR DEBTORS, *supra* note 1, at 66–68. Extrapolating from this data, it seems reasonable to conclude that debtors in dire financial straits—typically a condition that precedes any filing in bankruptcy—also have considerably less property than the general population of consumers in this country.

13. *See* DOUGLAS G. BAIRD & THOMAS H. JACKSON, CASES, PROBLEMS AND MATERIALS ON BANKRUPTCY 13–14 (2d ed. 1990) ("when the debtor is an *individual*, involuntary remedies are not available against all of the debtor's property")(emphasis added).

14. *See* ROCK, *supra* note 11, at 69–70.

15. *See* U.C.C. § 9–503 (1989).

16. A credit seller, who does not take a security interest under the provisions of Article Nine of the Uniform Commercial Code, *id.* §§ 9–101 - 9–507, is typically only a general creditor. He is given no better claim to the goods that he sold than any other general creditor, and he may acquire a lien on the goods, as on any other non-exempt property of the debtor, only by judicial action to enforce his claim. One limited exception exists in the Sales Article of the Uniform Commercial Code:

> Where the seller discovers that the buyer has received goods on credit while insolvent he may reclaim the goods upon demand made within ten days after the receipt, but if misrepresentation of solvency has been made to the particular seller in writing within three months before delivery, the ten day limitation does

debtor's perception that they have self-help rights by vague references to taking assets, without detailing when and how this may occur. Whether such references violate laws regulating non-judicial collection practices will often be problematic in the relevant jurisdiction.[17] But even when a particular collection practice constitutes a clear violation of applicable law, transaction-cost hurdles and evidentiary problems of enforcement abound.

Debtors' misperceptions of ways in which their property may be taken by creditors are not limited to overly broad applications of self-help remedies. Many debtors will likely fail to distinguish between secured and unsecured creditors' rights to use judicial remedies to effect seizure of property prior to judgment. While secured creditors in many jurisdictions can resort to judicial remedies that regularly permit the property that is their collateral to be taken prior to notice and an opportunity for the debtor to be heard in a court proceeding,[18] seizure of any of the debtor's property by unsecured creditors must generally await judgment and the issuance of a writ of execution.[19] A final misperception by debtors of creditors' rights to take their property arises from debtors' inadequate knowledge of the execution process. Few debtors are probably aware of the significant limitations that exemption laws place on levies under writs of execution.[20]

The unsophisticated debtor's worse scenario of being taken to court results from his perception of creditors' remedies as virtually boundless. Debtors' misperceptions do more, however, than bolster the effectiveness of informal collection practices *vis-a-vis* judicial ones. They have another significant impact on coercive collection. A debtor's misperceptions of creditors' remedies may open the priority contest played in the self-help arena to creditors who would have no chance of successfully competing if the

---

not apply. Except as provided in this subsection the seller may not base a right to reclaim goods on the buyer's fraudulent or innocent misrepresentation of solvency or of intent to pay.

Id. § 2–702(2). The few credit sellers who qualify for this statutory reclamation right, a type of lien arising by operation of law, may have access, like Article Nine secured parties, to remedies like replevin that provide for a taking prior to final judgment, a remedy typically not available to general creditors. But the failure of § 2–702(2) to provide for repossession without judicial process, a remedy expressly given Article Nine secured parties by U.C.C. § 9–503, casts serious doubt on the self-help repossession rights of a seller proceeding solely under § 2–702(2). *Compare id.* § 2–702(2) *with id.* § 9–503.

17. *See* ROBERT J. HOBBS, NATIONAL CONSUMER LAW CENTER, FAIR DEBT COLLECTION 423–28 [hereinafter NATIONAL CONSUMER LAW CENTER].

18. *See* JAMES J. WHITE & ROBERT S. SUMMERS, UNIFORM COMMERCIAL CODE 1206–07 (3d ed. 1988).

19. *See* EPSTEIN, *supra* note 4, at 18–20.

20. *See* STEFAN A. RIESENFELD, CASES AND MATERIALS ON CREDITORS' REMEDIES AND DEBTORS' PROTECTION 289–333 (4th ed. 1987).

contest occurred in court. Competition in self-help contests thus may be more extensive and, because based on a war of words, more intense than competition in priority battles that are judicially determined.

When the priority contest occurs in the forum of the debtor's mind, as it necessarily must when judicial process and self-help repossession or retention of the debtor's property by the creditor are absent, no creditors, regardless of their rank under formal rules of law, need concede defeat. Even creditors with liens that do sanction immediate seizure of a defaulting debtor's property, will often limit the use of their liens to bolstering the credibility of their demands to be paid before other creditors in those many instances in which the cost of taking and selling the encumbered property approaches or exceeds its forced sale value. Observers of collection practice know that a secured creditor may visit the debtor to impress him with the creditor's usually feigned interest in the debtor's household goods.[21] But in many cases, much of the debtor's property is not held hostage by a secured creditor at the time the contest among creditors begins in earnest, so there is, aside from the debtor's misperceptions about who can claim what and how, something for creditors to fight over.

Both economics and law account for the presence of prizes for successfully competing creditors to capture even after the debtor has become seriously overextended. Creditors frequently fail to take security interests in a debtor's property because of cost constraints that limit the use of collateral to secure small debts. Even the use of form contracts and self-help repossession is not costless. Additionally, federal law now proscribes any creditor from taking either a security interest in wages or other earnings[22] or a non-possessory, non-purchase money security interest in household goods.[23] Paradoxically, one aspect of these laws designed to protect debtors tends to harm them by insuring that property valued by the debtor may be up for grabs and therefore provide a pressure-generating device for cred-

---

21. *See* NATIONAL CONSUMER LAW CENTER, *supra* note 17, at 16.
22. FTC Trade Regulation Rule, 16 C.F.R. § 444 (1996). The rule defines "earnings" as "[c]ompensation paid or payable to an individual or for his or her account for personal services rendered or to be rendered by him or her, whether denominated as wages, salary, commission, bonus, or otherwise, including periodic payments pursuant to a pension, retirement, or disability program." 16 C.F.R. § 444.1(h) (1996).
23. *Id.* § 444.2(a)(4). "Household goods" is defined by the rule as:
   [c]lothing, furniture, appliances, one radio and one television, linens, china, crockery, kitchenware, and personal effects (including wedding rings) of the consumer and his or her dependents, provided that the following are not included within the scope of the term 'household goods': (1) Works of art; (2) Electronic entertainment equipment (except one television and one radio); (3) Items acquired as antiques; and (4) Jewelry (except wedding rings).
Id. § 444.1(i).

itors in the self-help collection contest. The rule proscribing a security interest in wages encompasses the debtor's human capital, usually an individual's most valuable asset, while that proscribing a security interest in household goods often covers items of sentimental or idiosyncratic value to the debtor. Both of these rules provide creditors with a source of leverage at the time a collection contest develops.

Although exemption laws protect some or all wages and often protect household goods from forced sale under judicial process initiated by unsecured creditors,[24] this fact only has significance in a war of nerves if the debtor is cognizant of the exemptions. So long as the debtor has access to some or all of his wages, unsecured creditors are not powerless in the informal or self-help collection process. And the debtor will have such access, for while state exemption laws vary widely, federal law provides a minimal standard of protection against the taking of wages by the judicial process of garnishment.

In an out-of-court contest, the measure of one creditor's advantage over another is directly related to the sophistication of the debtor. In the debtor's mind, persistence in demands for payment may elevate a creditor of only mean status before the law to a much higher rank. Because verbal collection contests offer such vast opportunities to establish and reorder priority, the forces they unleash upon the debtor may greatly exceed those associated with the actual seizure of property by judicial process.

Both the pervasiveness of informal collection[25] and its potential for harm have not escaped scrutiny by lawmakers. Of course, debt collectors have never been immune from common law tort actions such as assault, trespass, slander, intentional infliction of emotional distress and invasion of the right of privacy.[26] Moreover, in exercising its power to prohibit "unfair or deceptive acts or practices," the Federal Trade Commission has brought administrative actions against debt collectors for years.[27] Tort law

---

24. While the law of exemptions does not bar a creditor who has acquired a lien by operation of law or by the consent of the debtor from sale of the encumbered property, exemption law does restrict the property that unsecured creditors may subject to judicial process. *See* JORDAN & WARREN, *supra* note 2, at 62–3. Here again, the unsecured creditor has reason to prefer vague references to legal action over actual use of that process.

25. *See* William C. Whitford, *A Critique of the Consumer Credit Collection System*, 1979 WIS. L. REV. 1047, 1048 (most collection of delinquent consumer debts results from voluntary payments following bargaining between debtor and creditor, but creditors' legal remedies play an important role in determining the outcomes of that bargaining).

26. These and other torts that have arisen in debt collection cases are discussed in NATIONAL CONSUMER LAW CENTER, *supra* note 17, at 215–18.

27. In 1968, the Federal Trade Commission codified and supplemented this case law by implementing Guides Against Debt Collection Deception, but the Commission removed the Guide in 1995. *See* former 16 C.F.R. Part 237 (1994) removal reported at 60 Fed. Reg. 40265 (1995).

has been further supplemented by the enactment of statutes in many states designed to specifically regulate the actions of collection agencies by prohibiting certain practices such as "using or threatening violence, communicating in the name of a lawyer, using forms that simulate legal process or names that simulate the names of governmental entities, and publishing deadbeat lists."[28] The scope of these laws were expanded in some states in the 1970's to cover not just the actions of collection agencies but those of creditors collecting their own claims as well. Perceiving action in some states to be inadequate, however, in 1978 Congress proscribed certain practices by third-party debt collectors in the Fair Debt Collection Practices Act.[29]

But while federal and state legislation and administrative law and the common law's traditional and evolving tort doctrines have addressed the more flagrant abuses, the function of the self-help collection process has not changed. The creditor must get the debtor to honor obligations for yesterday's acquisitions of goods and services over the desires and possibly even urgent needs of the moment, and to honor the creditor's claim over all others.[30]

There is ample evidence that creditors believe that the informal or nonjudicial collection process is their most cost-effective remedy. Virtually no institutional lender is without its collection department, and independent collection agencies supplement the work of these departments and serve the needs of smaller creditors. From the standpoint of sheer volume of activity, self-help remedies continue to occupy center stage in American consumer-credit collection practice. One empirical study found that only about 4 percent of the claims against debtors filing for bankruptcy had been the subject of lawsuits at the time those debtors sought bankruptcy relief, although these debtors were doubtlessly in default on many of their obligations long prior to that time.[31]

Reordering priorities in the debtor's mind rather than under rules of law may even be the primary reason for a creditor's actual use of legal process. The need to convince debtors of the creditor's tenacity provides greater impetus for some judicial actions than the value of any property likely

---

28. MICHAEL W. GREENFIELD, CONSUMER TRANSACTIONS 544 (2d ed. 1991).

29. *See id.* at 543–47.

30. *See* NATIONAL CONSUMER LAW CENTER, *supra* note 17, at 16 ("Reasonably, the objective of a debt collector is, first, to identify the relatively few delinquent consumers who can afford to repay their debts and to convince them to do so and, second, to convince consumers who are insolvent to pay this collector rather than (or before) another creditor or expense.").

31. SULLIVAN ET AL, AS WE FORGIVE OUR DEBTORS, *supra* note 1, at 305.

to be seized by use of a judicial writ.[32] Moreover, while the bank or finance company that levies on sufficient property to satisfy its claim against a borrower need not concern itself further with that individual's perception of its tenacity, that lender derives an additional benefit from successful use of legal process by enhancing its reputation with other borrowers as one not to be trifled with.[33] Instances in which projection of this image tip the scales in favor of bringing an economically marginal action are not uncommon.[34] Recognition of the role of creditor reputation in the collection process reveals therefore that all legal process employed by creditors who have numerous debtors functions to some extent as part of a larger self-help strategy.

(4) Beneficial Aspects of Rewarding Individual Creditor Diligence

Having examined in Kelly's case in the preceding chapter the potential for harmful diligence engendered by competition among creditors, this work must now distinguish a second aspect of creditor diligence that justifies, on balance at least, the policy of rewarding individual creditors for their selfish actions. Creditor diligence is rewarded for the most fundamental of economic reasons: productivity in collection reduces the costs of credit.[35] This economic benefit occurs when the diligent creditor's recovery increases the aggregate recovery of all the debtor's creditors. Either the diligent creditor effects recovery from assets that the debtor would

---

32. *See* HERBERT JACOB, DEBTORS IN COURT: THE CONSUMPTION OF GOVERNMENT SERVICES 98–104 (1969) (finding that garnishment of wages usually produced settlements through out-of-court negotiations, which in a substantial number of the cases studied transferred the debtor's delinquency from one creditor to another).
33. *See* ROCK, *supra* note 11, at 112.
34. *Id.*
35. The principle of rewarding individual creditor diligence that benefits the debtor and other creditors of the debtor is not limited to creditor action that occurs only after the debtor defaults. One commentator also advances creditor diligence as a justification for the priorities accorded secured creditors who obtain a consensual lien on their debtors' principal assets at the time of entering into a contract to extend credit. While the secured creditor's gain would appear to be offset by depletion of the pool of unencumbered assets available to general creditors, Robert Scott argues that secured credit need not result in only a zero-sum game. He observes that the leverage provided by taking the debtor's assets hostage by means of a security interest enables the secured creditor to influence business decisionmaking for the good of all parties, including unsecured creditors, with an interest in the venture. *See* Robert E. Scott, *A Relational Theory of Secured Financing*, 86 COLUM. L. REV. 901, 904–33 (1986).

otherwise have withheld from all creditors, or that creditor alters the debtor's financial course for his benefit and that of one or more other creditors as well. Here, the selfish but diligent creditor is, in the memorable words of Adam Smith, "led by an invisible hand to promote an end which was no part of his intention."[36]

If the race-of-diligence rule were not capable of producing such benefits, history would doubtlessly have abandoned it in its primal stages, if, indeed, it would have ever formulated it. Instead, a rule requiring cooperation and sharing by all creditors would exist in its place. Bankruptcy law, historical and modern, recognizes the need for some form of collective execution, but that need is premised, according to a leading authority on early bankruptcy, on two necessary antecedents: "(a) insolvency, actual or apparent, of the debtor, and (b) plurality, actual or potential, of the creditors."[37] Why should the law impose even informal collective action with its attendant costs in those many instances when creditors can satisfy their claims through more cost-effective individual recovery without danger of aggregate creditor loss?[38]

Certainly *ad hoc* creditor action makes more sense than collective action when a debtor has defaulted on only one obligation for a reason, such as misunderstanding, complaint, or temporary cash shortage, that need not concern all that debtor's creditors. In the initial stage of a creditor's effort to collect on an overdue claim, the creditor is probably more concerned with strengthening the debtor's resolve to pay his claim than he is with establishing his priority over the debtor's other creditors.[39] Only when his initial collection efforts fail does the creditor perceive that he may be engaged in a contest with other creditors for resources that may prove inadequate to pay all creditors' claims.

(5) The Incidence of Harmful Aspects of the Race of Diligence

If the actions of individual creditors to collect their claims may have no effect or even a salutary one on the recoveries of the debtor's other credi-

---

36. 4 ADAM SMITH, AN INQUIRY INTO THE NATURE AND CAUSES OF THE WEALTH OF NATIONS 423 (Edwin Cannan ed. 1937) (1776).

37. Louis E. Levinthal, *The Early History of Bankruptcy Law*, 66 U. PA. L. REV. 223, 225–26 (1918).

38. *Id.* at 226. ("If the debtor has enough assets to meet all his debts, there is no need to seek special regulation to protect the creditors from one another. Each creditor may proceed individually against the debtor's property without in any way jeopardizing the chances of the other creditors of obtaining satisfaction of their claims, and the principle of priority can safely and properly be allowed to control.").

39. *See supra* note 6.

tors, that is certainly not always the case. Although creditors are often aided by Smith's invisible hand, in other instances they are "injured by the malevolent back of that hand."[40] Kelly's case, reported in the preceding chapter, clearly shows that in some instances creditors so over indulge in coercive collection effort that they precipitate bankruptcy or other financial failure to the embarrassment of their aggregate recovery.

Based on my observations of the consumer-credit industry over more than 30 years, I believe that failure to grant extensions in needed and promising cases has been widespread. At various times during this period, I have served in different roles—law teacher, practitioner, consultant, lecturer, writer, and member of the board of directors of a credit union—that have acquainted me with the practices employed in collecting consumer debt. While I know that there are numerous cases in which creditors do work with debtors by extending payment terms even when other creditors of those debtors do not, I believe that, unless the factor of competition among creditors is controlled, it significantly restricts the number of cases in which this occurs. The lone cooperating creditor is simply too vulnerable in many cases to strategic behavior by the debtor's other creditors.

Moreover, a creditor, who assumes, for selfish or humanitarian reasons, a disproportionate share of the risks of the workout, is sometimes penalized by courts who apparently fail to recognize that the creditor's actions may be in the best interests of the debtor and other creditors. In *In re Matthews*,[41] the court held that the lender who had financed the debtors' acquisition of collateral lost its purchase-money status when it refinanced the loan to cure the debtors' default and extend the time for repayment. The loss of the lenders's purchase-money status permitted the debtors in their subsequent bankruptcy to set aside the security interest of the lender under a provision of the Bankruptcy Code that permits debtors to avoid non-purchase money security interests that impair their exemptions in certain property.[42] The loss of the finance company's purchase-money status was an unfortunate call based on a formal rather than a functional distinction, as the source of the refinanced indebtedness obviously remained the loan used to acquire the collateral. The legislative purpose in permitting debtors in bankruptcy to set aside blanket security interests in household and certain related goods was to forestall creditors' use of leverage resulting from threats to take goods that are perceived as being worth more

---

40. RUSSELL HARDIN, COLLECTIVE ACTION 6 (1982). Hardin applies his metaphor to the problem of securing beneficial collective action in all contexts, and does not specifically address it to creditors' failure to cooperate in an advantageous workout. The workout, however, is clearly a type of collective action. *See id.* at 6–9.
41. 724 F.2d 798 (9th Cir. 1984).
42. 11 U.S.C.A. § 522(f)(1)(b)(i) (West Supp. 1996).

to the debtor from a replacement standpoint than to the creditor from a resale one. Whether that purpose, part of a "lost-value premise" for restricting creditors' remedies that will be examined in the last chapter, has merit or not, it does not justify avoiding a lender's security interest in specific property, the acquisition of which he has recently financed and which is apt to represent more than just leverage value to him.

My personal observations on the vulnerability of the lone cooperating creditor are bolstered by examining the shortcomings of traditional means of obtaining creditor cooperation in the next section of this chapter and by an analysis in the following chapter that applies decision-making theory to further explore the reasons for counterproductive coercive collection practices. Finally, I support my personal observations of collection practices with evidence of the rapid growth of Consumer Credit Counseling Services, an agency that later analysis will show to be particularly well qualified to deal with the problems of harmful coercive actions in the collection of consumer debt.

I am aware of no studies that would shed scholarly light on the extent of coercive collection that is counterproductive to creditors, and there may be many difficulties in conducting such empirical research. The question of counterproductive coercive collection raises the related one of whether a substantial number of debtors in bankruptcy would have the ability to make significant payments absent bankruptcy relief, for debtors with good incomes who could effect workouts in whole or in part may file for bankruptcy instead of attempting workouts simply to avoid coercive collection practices. These two related questions differ, however, for doubtlessly some debtors seek bankruptcy relief as an easy way out of their financial problems even when not driven to do so by the coercive collection practices of their creditors. Moreover, research on the issue of the number of debtors who could have made substantial payments had they not chosen bankruptcy is inconclusive. The two empirical studies that evaluate the ability of debtors in bankruptcy to repay significant parts of their obligations reach significantly different conclusions.[43]

Regardless of the difficulty of measuring the extent of counterproductive coercive collection, it appears that until recent times and the advent of Consumer Credit Counseling agencies, the practice was widespread in this country. Creditors contributed to the undoing of their consumer debtors by refusing pleas for forbearance in the collection of their claims, and because creditors did so time and time again, the problem assumed systematic proportions.

---

43. *Compare* SULLIVAN ET AL. *supra* note 1, at 205–19 (only a handful of debtors in bankruptcy could pay their debts by making payments from future income) *with* PURDUE STUDY, *supra* note 1, at 90 ($1.1 billion dollars of debt was discharged in Chapter 7 bankruptcies that could have been repaid out of future income).

The concluding section of this chapter examines the methods that have evolved over centuries of debtor-creditor law and practice for remedying the problem of counterproductive competition among creditors. It pays particular heed to the shortcomings of these traditional methods, particularly when they are applied to consumer credit, an industry that was still in its adolescence in this country just a half-century ago. Only by examining the deficiencies in a consumer-credit context of methods that have long been used with some degree of success in commercial cases can one appreciate the need for different solutions to the problem of overly coercive collection in consumer cases.

B. Models of Creditor Cooperation Based on Bankruptcy and Contract Law

(1) Introduction

The law has long recognized the merits of cooperation when collective action is invoked by bankruptcy or simply by agreement of creditors to forego the race of diligence. In the latter instance, courts had no difficulty in finding consideration and thus an enforceable agreement in the promises of two or more creditors of a common debtor to take less than the amount to which each was entitled or to extend the time for payment of their claims, whether reduced in amount or not. For centuries then these contracts, known respectively as composition and extension agreements, have provided a basis for addressing the harms of overly coercive collection, and they have coexisted over much of that period with a collective judicial remedy providing like relief. The very genesis of bankruptcy law was concern for the plight of creditors.[44] Although contemporary bankruptcy reflects a fundamental, often conflicting concern for the debtor, there is still no more basic principle in bankruptcy than that of maximizing the aggregate recovery of creditors.[45] Creditors' interests in bankruptcy are furthered either by orderly liquidations that generate greater value than piecemeal sales[46] or by reorganizations, debt adjustments and rehabilitations that preserve going-concern values that may exceed liquidation ones.[47]

---

44. *See* Levinthal, *supra* note 37.
45. *Id.*
46. *See* BAIRD & JACKSON, supra note 13, at 40 (when creditors act individually, assets may be sold off for less than if a single individual had controlled their disposition).
47. *Id.* (one creditor's levy on a key asset may lead to a cessation of the debtor's productivity to the detriment of all other creditors).

Moreover, in bankruptcy proceedings as well as under the common law of contract, the law encourages creditors to cooperate among themselves and with the debtor by according legal recognition to their agreements.[48]

Bankruptcy and contract-law models for securing beneficial cooperation are not to be lightly regarded, for they often furnish satisfactory, even optimal solutions to the problems of debtor overextension. The analysis that follows first examines how bankruptcy and contract law may further beneficial cooperation among creditors but then explores the inadequacies of these means of protecting creditors' and debtors' interests in many consumer credit cases.

(2) Developing the Bankruptcy Model: The Automatic Stay and the Treatment of Preferences and Setoffs in Bankruptcy

The automatic stay of any action by individual creditors to collect their claims other than by proceedings in the bankruptcy court, which is imposed upon the filing of a petition in bankruptcy,[49] is a clear recognition by lawmakers of the benefits that may be derived from substituting collective for individual action.[50] Obviously the bankruptcy court's duty to administer the estate for the benefit of all creditors would be thwarted if individual creditors could continue their self-serving ways outside that court once its jurisdiction was invoked. To insure that they do not and that the

---

48. *See* EPSTEIN, *supra* note 4, at 133–36. Common law composition and extension agreements are not binding on non-assenting creditors. In bankruptcy proceedings, the required number of a class of creditors with the required amount of claims may bind all members of that class by their acceptance of a plan of reorganization. *See* 11 U.S.C. §§ 1126(c), 1129(a)(8)(A) (1994).

49. 11 U.S.C.A. § 362 (West 1993 & Supp. 1996).

50.
> The automatic stay is one of the fundamental debtor protections provided by the bankruptcy laws. It gives the debtor a breathing spell from his creditors. It stops all collection efforts, all harassment, and all foreclosure actions. It permits the debtor to attempt a repayment or reorganization plan, or simply to be relieved of the financial pressures that drove him into bankruptcy.
>
> The automatic stay also provides creditor protection. Without it, certain creditors would be able to pursue their own remedies against the debtor's property. Those who acted first would obtain payment of their claims in preference to and to the detriment of other creditors. Bankruptcy is designed to provide an orderly liquidation procedure under which all creditors are treated equally. A race of diligence by creditors for the debtor's assets prevents that.

H.R. REP. No. 595, 95th Cong., 2d Sess. 340 (1978), *reprinted in* 1978 U.S.C.C.A.N. 5963, 6296–97.

opportunity for rehabilitation or orderly liquidation exists, the automatic stay is broadly drafted to proscribe self-help efforts[51] as well as judicial actions[52] by both secured and unsecured creditors.

Further evidence of the need to check harmful aspects of creditor competition is found in provisions of bankruptcy law that cancel creditor gain from certain transactions that occur even before a bankruptcy proceeding is commenced. In certain instances, creditors who receive pre-bankruptcy payment of their claims or who obtain pre-bankruptcy liens that secure previously unsecured claims will see those payments or liens avoided by the debtor's subsequently appointed trustee in bankruptcy as preferential transfers. Likewise, in certain instances, a creditor who setoffs his obligation to the debtor against the debtor's obligation to him may have his setoff set aside in a subsequent bankruptcy proceeding of the debtor.

Aspects of the bankruptcy law of preference and setoff are examined below, not because, as we shall later see, this law sees extensive use in consumer bankruptcies. Laws concerning preferences and setoffs are examined to show Congressional concern with a point of central concern to me—the need to address the harm that may result from unbridled creditor competition during the debtor's slide into bankruptcy. The further analysis of the different treatment accorded preference and setoff in analogous transactions again reflects Congressional concern with securing creditors' cooperation in promising workouts once the debtor's plight makes creditors' concessions necessary. It does so by focusing on a fine distinction drawn by Congress in providing greater incentives for securing that cooperation in the case of a creditor entitled to setoff than one who is having his claim to certain collateral subjected to the test of preferential transfer law. The examination of those aspects of bankruptcy law that follows is also necessary to understand the fundamentals of preference and setoff as a first step in ascertaining why these traditional devices for deterring or remedying some of the harmful aspects of creditor competition are of little effect in consumer-bankruptcy cases.

Provisions for the recovery of preferences[53] bespeak bankruptcy law's concern with providing some measure of protection against predatory collection practices that occur before bankruptcy but during the period in which the debtor's insolvency[54] is most apt to compound the damages to aggregate creditor recovery inflicted by individual collection effort. University

---

51. *See* 11 U.S.C.A. § 362(a)(3)-(7) (West 1993 & Supp. 1996).
52. *See id.* § 362(a)(1)-(8).
53. *See* 11 U.S.C. A. § 547 (West 1993 & Supp. 1996).
54. *See id.* § 547(b)(3) (imposing the requirement on the trustee's preferential transfer avoiding power that the debtor be insolvent at the time the property is transferred).

of Chicago law dean Douglas Baird and University of Rochester president Thomas Jackson employ the concept of the common pool to illustrate the need for collective action to curtail waste once a debtor is insolvent.[55]

So long as the aggregate catch of a few individuals fishing in a small lake is limited relative to the number of fish therein, their aggregate catch does not deplete the lake of sufficient fish to reproduce and maintain the population for future years. Like the prudent fishing of that lake, the individual actions of creditors of a solvent debtor are not apt to commit waste. Where the debtor has sufficient assets to satisfy all claims, no creditor has an economic incentive to dip too deeply into the debtor's pool of assets and diminish her ability to generate future income. Even if the debtor pays one creditor only after all others have been paid, the debtor possesses sufficient assets to make that last payment.

But the present demand for fish from the lake and for payment from the debtor may increase relative to the present supply of assets available to satisfy those demands. Just as too many fishers after too few fish will disproportionately sacrifice future for present value, too many creditors chasing the too few assets of an insolvent debtor will do likewise. In both cases, an individual taking from the common pool may know that her actions destroy-long-term values. But without some means of obtaining the cooperation of others, she will rationally continue to get as much as she can before the supply is exhausted. When a significant number of creditors of an insolvent debtor take these actions, a piecemeal liquidation of assets will result in which those assets yield less value than their worth in either an orderly liquidation or a successful rehabilitation of the debtor.

Preference law recognizes when recovery of claims by individual creditors is most likely to be harmful to creditors' aggregate recovery. A transfer of the debtor's property to a creditor to pay or secure an antecedent debt is only recoverable from that creditor as a preference if the debtor is insolvent at the time of the transfer. Hence, insolvency equates with the point at which the pond is fished beyond the ability of the fish remaining to produce sufficient offspring to restock it.

The law, however, does not avoid all pre-bankruptcy transfers of an insolvent debtor's property to creditors on account of antecedent debt as preferences. The preference section of the Bankruptcy Code imposes various other requirements on the use of that avoiding power, and also protects certain transfers that meet all those requirements from avoidance as preferences. Probably the most significant limitation on the power to avoid preferences is the requirement that the transfer occur within 90 days or,

---

55. *See* BAIRD & JACKSON, *supra* note 13, at 39–43; JACKSON, *supra* note 2, at 10–12; Thomas H. Jackson, *Avoiding Powers in Bankruptcy*, 36 STAN. L. REV. 725, 727–31, 756–68 (1984).

when the transferee has a close relationship with the debtor and is classified an "insider," within one year before the filing of the petition initiating the bankruptcy action.[56]

Avoidance of a transfer under the preference section gives the trustee the right to recover the property transferred, or its value, for inclusion in the bankruptcy estate. The trustee's power to avoid some transfers of the debtor's property that are indefeasible outside of bankruptcy proceedings enriches the bankruptcy estate,[57] enabling successful rehabilitation or reorganization to occur in some cases[58] and larger dividends to be paid creditors in others.

For this reason, the justification for recovery of a preference, even in a Chapter 7 bankruptcy liquidation, is not limited to the equitable one of treating like creditors alike. Where the property recovered brings more in an orderly liquidation in bankruptcy than the credit the debtor received for it in the pre-bankruptcy transfer, the avoidance is not merely a zero-sum transaction designed only to produce equality of treatment among creditors.[59]

The power to avoid preferences may enrich the pool of assets out of which creditors will be paid in other ways as well. By discouraging pursuit of spoils by individual creditors when bankruptcy seems imminent,[60] that avoiding power provides an incentive to creditors to initiate voluntary cooperative undertakings,[61] which may lead to the debtor's financial recov-

---

56. *See* 11 U.S.C.A. § 547 (West 1993 & Supp. 1996).

57. 11 U.S.C. § 541 (a)(3) (1994); 11 U.S.C.A. §§ 547, 550 (West 1993 & Supp. 1996).

58. *Cf.* United States v. Whiting Pools, Inc., 462 U.S. 198, 203 (1983) (The court determined that property on which the government levied under a tax lien one day before the debtor filed a Chapter 11 reorganization petition became part of the bankruptcy estate under an ambiguous provision of the Bankruptcy Code. The court relied upon policy arguments stressing the need of property for reorganization and the benefits that may result from reorganization.)

> By permitting reorganization, Congress anticipated that the business would continue to provide jobs, to satisfy creditors' claims, and to produce a return for its owners. Congress presumed that the assets of the debtor would be more valuable if used in a rehabilitated business than if 'sold for scrap'. The reorganization effort would have small chance of success, however, if property essential to running the business were excluded from the estate.

Id. (citations deleted).

59. *See*, John C. McCoid II, *Bankruptcy, Preferences, and Efficiency: An Expression of Doubt*, 67 VA. L. REV. 249, 259–61 (1981).

60. *See id.* at 261–68 (recognizing that "[p]reference law operates through recapture and deterrence," but seriously questioning its efficacy in either function).

61. *See* H.R. REP. No. 595, 95th Cong., 2d Sess. 177, *reprinted in* 1978 U.S.C.C.A.N. 5963, 6138 (The preference avoiding power discourages creditors "from racing to the courthouse to dismember the debtor during his slide into bankruptcy. The protection thus

ery or otherwise obviate the costs of a bankruptcy proceeding. Judge Easterbrook's explanation of the policy underlying preference law is instructive:

> The trustee's power to avoid preferences ... is essential to make the bankruptcy case a *collective* proceeding for the determination and payment of debts. Any individual creditor has a strong incentive to make off with the assets of a troubled firm, saving itself at potential damage to the value of the enterprise. Many a firm is worth more together than in pieces, and a spate of asset-grabbing by creditors could dissipate whatever firm-specific value the assets have. Like fishers in a common pool, creditors logically disregard the fact that their self-protection may diminish aggregate value—for if Creditor A does not lay claim to the assets, Creditor B will, and A will suffer for inaction. All creditors gain from a rule of law that induces each to hold back. The trustee's avoiding powers serve this end in two ways: first, they eliminate the benefit of attaching assets out of the ordinary course in the last 90 days before the filing, so that the rush to dismember a firm is not profitable from a creditor's perspective; second, the avoiding powers assure each creditor that if it refrains from acting, the pickings of anyone less civil will be fetched back into the pool.[62]

As Judge Easterbrook observed, because preference law recognizes the need to curtail predatory collection practices by creditors, that law takes into account how creditors obtain payment from their debtors in determining whether those payments are recoverable. While equitable distribution among creditors is[63] and has historically been[64] a major goal of preference law, not all transfers meeting a simple test of producing inequality of distribution if they be allowed to stand have been recoverable. The former Bankruptcy Act required proof that the creditor have reasonable cause to believe that the debtor was insolvent at the time of the transfer.[65] This requirement was not carried forward to the elements of a voidable preference contained in the present Bankruptcy Code. In the Bankruptcy Code, however, Congress has continued to reject the invalidation of transfers solely on the basis of inequality of distribution among creditors. Certain payments "made in the ordinary course of business or financial affairs of

---

afforded the debtor often enables him to work his way out of a difficult financial situation through cooperation with all of his creditors.").

62. Levit v. Ingersoll Rand Fin. Corp. 874 F.2d 1186, 1194 (7th Cir. 1989) (emphasis by the court) (citation deleted).

63. *See* H.R. REP. No. 595, 95th Cong., 2d Sess. 177–78, *reprinted in* 1978 U.S.C.C.A.N. 5963, 6138 ("the preference provisions facilitate the prime bankruptcy policy of equality of distribution among creditors of the debtor").

64. *See* Worseley v. Demattos, 96 Eng. Rep. 1160, 1161 (K.B. 1758); BAIRD & JACKSON, *supra* note 13, at 422–25.

65. *See* Bankruptcy Act § 60(b) (repealed 1978); 11 U.S.C. § 96(b).

the debtor and transferee" are immunized from attack as preferences,[66] even when made in the prescribed period preceding bankruptcy and by an insolvent debtor.

Legislative history reveals that "[t]he purpose of this [ordinary-course] exception is to leave undisturbed normal financial relations because it does not detract from the general policy of the preference section to discourage unusual action by the debtor or his creditor during the debtor's slide into bankruptcy."[67] In light of this legislative history, one court has held that the ordinary-course exception "should protect those payments which do not result from 'unusual' debt collection or payment practices."[68] Although the preference avoidance power takes its very name from the advantage the creditor receiving the transfer would gain over other creditors of her class in a bankruptcy proceeding if she were allowed to retain that transfer, the means a creditor employs to obtain a payment is no less a statutory requirement for its avoidance than the distributive end it serves.

A line of cases decided in recent years firmly establishes the focus of preference law on curtailing harmful collection practices. Because the exception that immunizes ordinary-course payments from avoidance as preferences infringes upon the other policy of preference law—that of equality of distribution among creditors of the same class in bankruptcy—the Ninth Circuit in a case decided in 1990 construed the ordinary-course exception narrowly. In *Matter of CHG International, Inc.*,[69] that court held that the ordinary-course exception does not cover payments on long-term debt, as distinguished from payments made on such obligations as trade credit or commercial paper. But the Sixth Circuit in *In re Finn*,[70] reached the opposite result. Noteworthy in appraising the relative merits of the two courts' decisions is that part of the legislative history contained in the

---

66.
(c) The trustee may not avoid under this section a transfer—...
(2) to the extent that such transfer was—
(A) in payment of a debt incurred by the debtor in the ordinary course of business or financial affairs of the debtor and the transferee;
(B) made in the ordinary course of business or financial affairs of the debtor and the transferee; and
(C) made according to ordinary business terms;
11 U.S.C. § 547(c)(2) (1994).
67. H.R. REP. No. 595, 95th Cong. 2d Sess. 373–74 (1978), *reprinted in* 1978 U.S.C.C.A.N. 5787, 6329.
68. Marathon Oil Co. v. Flatau (*In re* Craig Oil Co.), 785 F.2d 1563, 1566 (11th Cir. 1986). *See also* Waldschmidt v. Ranier (*In re* Fulghum Construction Corp.), 872 F.2d 739, 744 (6th Cir. 1989) (applying the test in *Craig*).
69. 897 F.2d 1479 (9th Cir. 1990).
70. Gosch v. Burns (*In re* Finn), 909 F.2d 903 (6th Cir. 1990).

House Report of the Bankruptcy Code that addresses preference law's concern with both policies: promoting equality of distribution and discouraging coercive collection practices.

> The purpose of the preference section is two-fold. First, by permitting the trustee to avoid pre-bankruptcy transfers that occur within a short period before bankruptcy, creditors are discouraged from racing to the courthouse to dismember the debtor during his slide into bankruptcy. The protection thus afforded the debtor often enables him to work his way out of a difficult financial situation through cooperation with all of his creditors. Second, and more important, the preference provisions facilitate the prime bankruptcy policy of equality of distribution among creditors of the debtor. Any creditor that received a greater payment than others of his class is required to disgorge so that all may share equally. The operation of the preference section to deter "the race of diligence" of creditors to dismember the debtor furthers the second goal of the preference section—that of equality of distribution.[71]

While the Ninth Circuit in *CHG International* noted that discouraging a race to collect is a "second policy" underlying preference law, it observed that the "foremost purpose behind the voidable preference provision is to assure fair or equal treatment of all creditors within the same class."[72] Simply subordinating one of the dual purposes of preference law to the other may be superficially appealing in cases in which they appear to conflict. Subordination, however, does not seriously address either the comment in the legislative history or the statutory text of the ordinary-course exception. A better method would attempt to accommodate both purposes, and a subsequent decision by the Ninth Circuit, which was based on its ruling in *CHG International,* failed the test of appeal in the Supreme Court.[73]

In ruling that the ordinary-course exception does apply to payments on long-term debt, the Supreme Court in *Union Bank v. Wolas* addressed the seeming conflict between the two policies of preference law set forth in the legislative history and rejected the reasoning of the Ninth Circuit.

> As this comment demonstrates, the two policies are not entirely independent. On the one hand, any exception for a payment on account of an antecedent debt tends to favor the payee over other creditors and therefore may conflict with the policy of equal treatment. On the other hand, the ordinary course of business exception may benefit all creditors by deterring

---

71. H.R. Rep. No. 595, *supra* note 63.
72. 897 F.2d at 1482–83.
73. *In re. ZZZZ Best Co., Inc.*, 921 F.2d 968 (9th Cir. 1990), *rev'd sub nom.* Union Bank v. Wolas, 502 U.S. 151 (1991).

the "race to the courthouse" and enabling the struggling debtor to continue operating its business.[74]

The meaning of the last sentence may not be readily apparent. The policy of deterring a race to the courthouse may seem best served by use of the preferential transfer avoiding power and not by application of the ordinary-course exception to that avoiding power. But that will not be the case when creditors' fear of being unable to retain ordinary-course payments leads to the onslaught of a race of diligence that otherwise could have been avoided.

While the Supreme Court in *Union Bank v. Wolas* grounded its decision on the absence of any distinction in the treatment of long and short term debt in the text of the statute containing the ordinary-course exception, it further observed that the policy of discouraging action by creditors to dismember the debtor applied to both types of claimants and concluded:

> Thus, even if we accept the Court of Appeals' conclusion that the availability of the ordinary business exception to long-term creditors does not directly further the policy of equal treatment, we must recognize that it does further the policy of deterring the race to the courthouse and, as the House Report recognized, may indirectly further the goal of equal distribution as well.[75]

By reconciling the two policies in this manner, the Supreme Court effected not only a literal application of the text of the statute but also recognized the provision in the legislative history, which I have previously cited, that expressly addresses the role of the ordinary-course exception. "The purpose of this exception is to leave undisturbed normal financial relations, because it does not detract from the general policy of the preference section to discourage unusual action by either the debtor or his creditors during the debtor's slide into bankruptcy."[76] The Court noted that this comment "clearly supports" the application of the ordinary-course exception to payments on long-term debt, and characterized the provision of the House Report that addressed the two-fold purpose of the preference section as "not inconsistent" with that position.[77] The Supreme Court's decision in *Union Bank v. Wolas* clearly underlines the important role the law of pref-

---

74. 502 U.S. at 161.
75. Id. at 162.
76. H.R. REP. No. 595, *supra* note 67.
77. Douglas Baird and Thomas Jackson have traced the troubled history of the policy underlying voidable preferences, exploring changes in both the role and meaning of the ordinary-course exception in that context. *See* BAIRD & JACKSON, *supra* note 13, at 422–27.

erential transfer is designed to play in preventing collection practices that tend to diminish aggregate creditor recovery.

Provisions regulating creditors' setoff rights[78] also evidence bankruptcy law's concern with providing incentives for securing individual creditor cooperation before the filing of a bankruptcy petition. A creditor who may setoff the amount she owes the debtor against the amount the debtor owes her protects herself in the same manner as does the holder of an indefeasible lien on the debtor's property.[79] To the extent of the creditor's indebtedness to the debtor, the creditor's claim is protected against the reduced payments usually received by creditors without either liens or setoff rights in bankruptcy.

Like the law of preferences, the law regulating setoff rights in bankruptcy is concerned with the means the creditor has employed to obtain her rights. These provisions of setoff law are similar in effect to preference law's avoidance of payments that are made out of the ordinary course.[80] But the rules governing setoff offer an incentive for creditor cooperation in an instance that the rules governing preferences lack. This additional incentive is probably based on a perceived greater need for obtaining the cooperation of a creditor entitled to setoff than one who, subject to the limitations imposed by preference law, is entitled to foreclose on collateral.

---

78. 11 U.S.C.A. § 553 (West 1993 & Supp. 1996). Except for certain limitations imposed by the statute, § 553 recognizes in bankruptcy the right of a creditor under nonbankruptcy law "to offset a mutual debt owing by such creditor to the debtor that arose before the commencement of the case under this title against a claim of such creditor against the debtor that arose before the commencement of the case." 11 U.S.C.A. § 553(a) (West 1993 & Supp. 1996). Because § 553 does not incorporate in its limitations the rules of the statute governing preferences contained in § 547, the preference avoiding power plays no role in regulating setoffs. See BAIRD & JACKSON, *supra* note 13, at 511.

79. The Bankruptcy Code expressly recognizes the analogy between the protection given a creditor by a right of setoff and that given a lien creditor by her claim to the encumbered property by defining both as secured claims. See 11 U.S.C. § 506(a) (1994).

80.
  [Setoff rights are not affected in bankruptcy,] except to the extent that—
  (2) such claim was transferred, by an entity other than the debtor, to such creditor—
  (A) after the commencement of the case; or
  (B)(i) after 90 days before the date of the filing of the petition; and
  (ii) while the debtor was insolvent; or
  (3) the debt owed to the debtor by such creditor was incurred by such creditor—
  (A) after 90 days before the date of the filing of the petition;
  (B) while the debtor was insolvent; and
  (C) for the purpose of obtaining a right of setoff against the debtor.
Id. § 553(a)(2)-(3).

An examination of the application of preference law to the following type of secured transaction is necessary to understand this difference in preference and setoff rules. A commercial finance company that takes inventory and receivables as collateral for its loan to a firm engaged in marketing goods, suffers a loss of its security interest when inventory is sold to buyers in the ordinary course of business,[81] and when receivables are collected by the firm from customers who were not instructed to make payment directly to the finance company.[82] The finance company must necessarily authorize and even bless these transactions as they furnish the means for the firm to fulfill its loan obligations. Identifiable proceeds of the lost inventory and receivables, such as cash and additional inventory and receivables, to which the security interest will attach,[83] offer some degree of protection against diminution of the finance company's collateral. But insofar as the firm fails to make payment on the loan with the proceeds of the collateral, either because the loan agreement does not require it or the firm violates the terms of one that does, the finance company relies primarily on the firm acquiring new inventory from suppliers and new receivables from further credit sales. By the use of an after-acquired property clause in the security agreement, the finance company's security interest will attach to the new inventory and receivables even if they are not identifiable as proceeds of its original collateral.[84]

For purposes of preferential transfer law, however, transfers of the debtor's property, including the granting of security interests therein,[85] can-

---

81. *See* U.C.C. § 9–307(1) (1989). *See also id.* § 9–306(2) (excising security interest in collateral when debtor's sale is "authorized by the secured party in the security agreement or otherwise").

82. *See id.* § 9–318(3). The danger of the merchant misappropriating the payment from his customer to the disadvantage of the finance company is not present in that type of receivables financing in which the finance company notifies the merchant's customers to make payments directly to the finance company. The growth, however, of "non-notification," receivables financing attests its utility. *See* 1 GRANT GILMORE, SECURITY INTERESTS IN PERSONAL PROPERTY 132 (1965) (except for financing in the textile industry and a few other fields, present-day accounts receivable financing is done on a non-notification basis).

83. *See* U.C.C. § 9–306(2) (1989).

84. The security agreement will typically provide that the obligations covered by it are to be secured by after-acquired collateral. As the Uniform Commercial Code recognizes the general validity of such a contractual term, *see id.* § 9–204(1), the secured party will not have to execute a supplemental security agreement when the debtor acquires additional collateral.

85. " '[T]ransfer' means every mode, direct or indirect, absolute or conditional, voluntary or involuntary, of disposing of or parting with property or with an interest in property, including retention of title as a security interest and foreclosure of the debtor's equity of redemption;" 11 U.S.C.A. § 101(54) (West Supp. 1991).

not occur until the debtor has rights in the property transferred.[86] For this reason, the finance company's security interest in that inventory acquired and those receivables created by the merchant within a statutorily defined period preceding bankruptcy[87] could be vulnerable to attack as preferences

---

86. *See* 11 U.S.C. § 547(e)(3) (1994). The same result is apparently also dictated by application of the rules of the Uniform Commercial Code governing the "attachment" of a security interest. *See* U.C.C. § 9–203(1)(c) (1989) (a security interest is not enforceable and does not attach unless the debtor has rights in the collateral). The seemingly redundant provision in § 547, the preference section of the Bankruptcy Code, is designed to expressly overrule those cases decided prior to legislation specifically addressing the issue of after-acquired interests in inventory and receivables. These cases may have afforded secured parties too large a measure of protection. They recognized the filing of a financing statement, rather than the attachment of the security interest upon the debtor's acquisition of rights in the collateral, as the time of transfer for purposes of applying preference law. These cases therefore could be interpreted as imposing no restrictions whatsoever on improvements in position by the secured party during the period preceding bankruptcy. *See* DuBay v. Williams, 417 F.2d 1277 (9th Cir. 1969); Grain Merchants of Ind., Inc. v. Union Bank & Sav. Co., 408 F.2d 209 (7th Cir. 1969).

While the law clearly supports the application of § 547(e)(3) to the creation of security interests, vestiges of the opposing judicial rule may remain in other areas. *See* Riddervold v. Saratoga Hospital (*In re* Riddervold), 647 F.2d 342 (2d Cir. 1981) (garnishment of wages earned within the preferential transfer period was not voidable because the writ, upon which the garnishment was based, was served on the debtor's employer before the commencement of that period and left the debtor with no property to be subsequently transferred).

Even after the debtor has acquired rights in the collateral and the security interest has attached under the rules of U.C.C. § 9–203, the secured party may delay the time of transfer for purposes of applying the law of preferences by failing to establish the priority of his interest against the holder of a hypothetical judicial lien on or within 10 days after the date of attachment. *See* 11 U.S.C.A. § 547(e)(1)-(2) (West 1993 & Supp. 1996). State law requires the secured party to perfect the security interest to establish priority. *See* U.C.C. §§ 9–201, 9–301(1)(b) (1989). The UCC's priority provision, however, need not concern the secured party who has already filed a financing statement covering inventory and receivables. The Uniform Commercial Code's notice-filing system does not require additional filing for further acquisitions of collateral, provided the previously filed financing statement describes the after-acquired collateral. *See id.* § 9–402 and its Official Comment 2. Based on a prior filing, ordinarily completed by any secured party properly planning his transaction, a security interest in inventory and receivables subsequently acquired by the debtor will be perfected as soon as it attaches. *See id.* § 9–303(1).

87. The bankruptcy trustee's power to avoid transfers that otherwise meet the tests of a preference is typically limited to those that are made on or within 90 days of the filing of the petition initiating the bankruptcy action. *See* 11 U.S.C. § 547(b)(4)(A) (1994). Where the transferee is an "insider," however, a transfer made within one year before the filing of the petition may be avoided if the other elements of a preference are present. *See id.* § 547(b)(4)(B). "Insiders" are entities with a close relationship with the debtor. *See* 11 U.S.C. § 101(31) (1994).

had Congress not provided special dispensation for transactions of this nature.[88] But the protection afforded those financing on the security of inventory and receivables, based on their necessary reliance on after-acquired property, does not protect their interests against avoidance as preferences in all instances. Bankruptcy law may create a voidable preference in inventory and receivables collateral to the extent that the amount of the creditor's claim over the value of the collateral is less on the day the bankruptcy petition is filed than it was at a prescribed earlier date, usually 90 days prior to filing the bankruptcy petition.[89] Thus, the security interest in the inventory and receivables on hand on the day the bankruptcy petition is filed is a voidable preference to the extent that the secured creditor has improved its position by decreasing the amount of its undersecured claim between the earlier and later dates "to the prejudice of other creditors holding unsecured claims."[90]

Although preference law normally prohibits improvements in position by creditors with security interests in inventory and receivables during a certain period preceding bankruptcy, setoff law addresses improvements in position shortly before bankruptcy differently. A creditor otherwise entitled to recover on a claim against the debtor by setoff of the amount that creditor owes the debtor may have setoff rights reduced in bankruptcy by an analogous improvement-in-position test only if the creditor actually exercises setoff rights within 90 days before bankruptcy.[91] When the cred-

---

88.
(c) The trustee may not avoid under this section a transfer—
(5) that creates a perfected security interest in inventory or a receivable or the proceeds of either, except to the extent that the aggregate of all such transfers to the transferee caused a reduction, as of the date of the filing of the petition and to the prejudice of other creditors holding unsecured claims, of any amount by which the debt secured by such security interest exceeded the value of all security interests for such debt on the later of—
(A)(i) with respect to a transfer to which subsection (b)(4)(A) of this section applies, 90 days before the date of the filing of the petition; or
(ii) with respect to a transfer to which subsection (b)(4)(B) of this section applies, one year before the date of the filing of the petition; or
(B) the date on which new value was first given under the security agreement creating such security interest;
11 U.S.C. § 547(c)(5) (1994).
89. *Id.*
90. *Id.*
91.
(b)(1) Except with respect to a setoff of a kind described in [other sections of the code], if a creditor offsets a mutual debt owing to the debtor against a claim against the debtor on or within 90 days before the date of the filing of the peti-

itor does not exercise the right of setoff before bankruptcy, any decrease in the amount by which that creditor's obligation to the debtor is insufficient to cover the creditor's claim against the debtor, measured by comparing this insufficiency at some pre-bankruptcy date and the date of bankruptcy, will not diminish the creditor's post-bankruptcy setoff rights.[92]

This difference in the treatment accorded secured creditors who have not resorted to their default remedies on inventory and receivables prior to bankruptcy and creditors who have not exercised their rights to setoff prior to bankruptcy is doubtlessly designed to give the latter a greater incentive than the former to refrain from precipitous action prior to bankruptcy. The ease with which a bank, by simple bookkeeping entry, may setoff the balance of deposits in the debtor's checking account against the balance on the debtor's note,[93] a process much less cumbersome and costly than a secured creditor's resort to default proceedings to realize the value of inventory and receivables collateral, even under the self-help remedies of Article 9 of the Uniform Commercial Code,[94] was no doubt perceived by Congress as sufficient reason for offering the creditor entitled to setoff an added incentive to cooperate in the debtor's workout. The technical distinction Congress has drawn in the use of the setoff and preferential transfer avoiding powers reflects additionally a broader and more fundamental policy concern though and that is Congressional recognition of the

---

tion, then the trustee may recover from such creditor the amount so offset to the extent that any insufficiency on the date of such setoff is less than the insufficiency on the later of—
(A) 90 days before the filing of the petition; and
(B) the first date during the 90 days immediately preceding the date of the filing of the petition on which there is an insufficiency.
(2) In this subsection, 'insufficiency' means amount, if any, by which a claim against the debtor exceeds a mutual debt owing to the debtor by the holder of such claim.
11 U.S.C.A. § 553(b) (West 1993 & Supp. 1996).

92. *See* Exxon Corp. v. Compton Corp. (*In re* Compton Corp.), 22 B.R. 276 (Bankr. N.D. Tex. 1982).

93. "Set-off is the cancellation of cross-demands between two parties. The term commonly is used to cover both judicially supervised set-offs and automatic extinction of cross-demands." Luize E. Zubrow, *Integration of Deposit Account Financing into Article 9 of the Uniform Commercial Code: A Proposal for Legislative Reform*, 68 MINN. L. REV. 899, 901, n.3 (1984).

94. *See* U.C.C. §§ 9–502(1) (secured party's right to collect receivables from obligors and account debtors), 9–503 (secured party's right to take possession of collateral by self-help or judicial action), 9– 504(3) (secured party's right to dispose of collateral by public or private proceedings).

importance of striving to fine tune the benefits of creditor competition with those of creditor cooperation.

(3) Developing the Contract Model: Composition and Extension Agreements Among Creditors

The search for means of securing cooperative undertakings between a consumer debtor and her creditors need not be limited to those offered by the law of bankruptcy. Formal composition and extension agreements offer a closer parallel to the kind of relief Kelly vainly sought than the protections afforded by potential and actual bankruptcy liquidation proceedings. Such contractual undertakings among creditors—the former being an agreement to discharge their claims upon less than full payment and the latter being one to extend the time for repayment—have long been judicially enforced as exceptions to the rule in the classic case of *Foakes v. Beer*,[95] finding no consideration for one creditor's promise to discharge a liquidated and matured debt on part payment thereof.[96] The doctrine of that case also applied to a single creditor's promise to extend the time for payment where the debtor's promise to pay interest on arrearages imposed no greater obligation than that previously assumed.[97] The strong justifications for the exceptions enforcing composition and extension agreements among creditors—that the consideration for which each creditor bargains is found in the promise of the debtor to treat creditors alike and the promises of other creditors to forego their former rights[98]—constitute judicial recognition of the benefits that may be derived from cooperative collective action.

---

95. 9 App. Cas. 605 (H.L. 1884).
96. *See* GRANT GILMORE, THE DEATH OF CONTRACT (1974), *quoting* 1 SAMUEL WILLISTON, CONTRACTS § 120 (1920)). Professor Gilmore seriously questions the origins of the rule of *Foakes v. Beer*. *See id.* at 30–34. Exceptions to the rule of that case are discussed in VERN COUNTRYMAN, CASES AND MATERIALS ON DEBTOR AND CREDITOR, 181–82 (2d ed. 1974).
97. *See* COUNTRYMAN, *supra* note 96, at 181.
98. *See* RESTATEMENT OF CONTRACTS § 84 comment d (1932).

C. Inadequacies in the Application of the Contract and Bankruptcy Models in Consumer Credit Cases

(1) Comparing Commercial and Consumer Credit Cases

Creditors of commercial debtors frequently use composition and extension agreements to achieve advantageous workouts. The number of such arrangements is difficult to measure with any degree of precision for two reasons. Unlike bankruptcy records, records of composition and extension agreements have no prescribed filing offices and therefore cannot be found at a prescribed location. Furthermore, parties do not always reduce these agreements to formal written contract but may create tacit standby agreements in which each creditor takes no action to collect over the amount provided by the arrangement if other creditors do likewise.[99] The number of continuing-education seminars and practice guides devoted to the subject of workouts by troubled firms evinces, however, an interest in compositions and extensions that must be generated by the widespread use of these arrangements in the collection of commercial debt.

These same creditors of commercial debtors also frequently benefit from the various forms of bankruptcy relief, either as an inducement for contractual undertakings or directly when bargaining models fail. The workout of a commercial debtor may consist of complementary negotiations and bankruptcy relief. One group of experienced practitioners reports that commercial workouts often take the form initially of out-of-court negotiations and discussions and that in some instances bankruptcy proceedings will not be required. But Chapter 11 bankruptcy reorganization often provides the means to effectuate a plan, the basics of which have been negotiated before the petition in bankruptcy was filed. In those cases, "the Chapter 11 process is needed to bind the few reluctant parties and to take advantage of the remedies only Chapter 11 provides."[100]

---

99. *See* DONALD LEE ROME, ET AL., BUSINESS WORKOUTS MANUAL 3–27 (2d ed. 1992) [hereinafter BUSINESS WORKOUTS MANUAL].

100. *Id.* at 1–7 - 1–8. *See also* Richard F. Broude, *Cramdown and Chapter 11 of the Bankruptcy Code: The Settlement Imperative*, 39 BUS. LAW. 441 (1984) ("Chapter 11 is biased toward settlement by the parties in interest in the case"); *Leveraged Debtors Opt for New Out: Prepackaged Chapter 11*, COM. L. BULL, Nov.- Dec. 1990, at 8–10 ("[C]hapter 11 proceedings in which prior to filing, the debtor has created its plan and

Collection Practices   61

While creditor cooperation both in and out of bankruptcy is often forthcoming in commercial lending, creditors of consumers are far less likely to accept a workout proposed by the debtor outside the bankruptcy court or to eagerly embrace one thrust upon them by Chapter 13 of the Bankruptcy Code. Moreover, bankruptcy liquidation proceedings under Chapter 7 will commonly produce even smaller recoveries for consumer lenders than for commercial ones.

The typical Chapter 7 bankruptcy of a consumer results in a discharge,[101] which insulates the consumer's future income,[102] frequently his only valuable asset,[103] from creditors' claims.[104] For this reason, creditors of consumer debtors virtually never welcome voluntary bankruptcy nor invoke involuntary proceedings,[105] even when they have knowledge of preferences and setoffs that may be recovered for their benefit only in bankruptcy.

Creditors are not likely to ascribe much value to the use of preference and set-off law in typical Chapter 7 consumer cases, for they will generally

---

obtained the approval of the required number of creditors holding the required amount of claims to bind their class are becoming more common.").

101. *See e.g.* JORDAN & WARREN, *supra* note 2, at 31 ("an individual will normally receive a discharge from pre-bankruptcy debts"); DAVID T. STANLEY & MARJORIE GIRTH, BANKRUPTCY PROBLEM, PROCESS, REFORM 59 (1971) ("the debtor almost always receives a discharge"). The court will deny a Chapter 7 discharge to a debtor who has engaged in certain misconduct or received a previous discharge in certain other bankruptcy proceedings commenced within 6 years before the filing of the subsequent Chapter 7 petition. *See* 11 U.S.C. § 727(a) (1994). And when discharge is given, certain debts may be excepted from it. *See* 11 U.S.C.A. § 523(a) (West 1993 & Supp. 1996).

102. Although the debtor's right to future earnings under an existing employment contract may come within the scope of property of the bankruptcy estate under the Bankruptcy Code's inclusion of "all legal or equitable interests of the debtor in property as of the commencement of the case," 11 U.S.C. § 541(a)(1) (1994), subpart (6) of that statute and subsection expressly excludes from the bankruptcy estate "earnings from services performed by an individual debtor after the commencement of the case." *Id.* § 541(a)(6). Nor will a security interest in the debtor's future earnings pass through bankruptcy under the rule of § 506(d) of the Bankruptcy Code and Long v. Bullard, 117 U.S. 617 (1886). *See* 11 U.S.C. § 552 (1994); Local Loan Co. v. Hunt, 292 U.S. 234 (1934).

103. Sullivan, Warren and Westbrook's financial data on debtors in bankruptcy contain arithmetic means for annual family income, total assets and secured debt. While assets were valued at $29,355, debt secured by these assets was $23,034, leaving the debtors a net worth in this property of only $6,321. Annual family income was $15,779. *See* SULLIVAN ET AL., *supra* note 1, at 64, Table 4.1. Even if no consideration is given to the probability of the debtors' earnings increasing over their lifetimes, any reasonable calculation of the present value of their future earnings should produce a substantially greater amount than $6,321.

104. *See* 11 U.S.C. § 727(b) (1994).
105. *See id.* § 303.

perceive that the costs of the debtor's discharge in these proceedings exceeds any advantages that they may derive from the bankruptcy trustee's use of these avoiding powers. The bright promise of preference and setoff law imposing restraints on unbridled individual creditor effort against troubled consumers who may seek bankruptcy relief is largely illusory for other reasons as well. Consumers frequently fail to formalize their financial demise by invoking bankruptcy, electing instead, like Kelly, to distance themselves in other ways from their creditors. Recall that Kelly moved without giving her creditors a forwarding address. Debtors may also distance themselves from their creditors by simply stonewalling their demands for payment. The debtor may succeed with this method when the creditor's projected costs of recovery exceed the amount of the claim discounted by the uncertainty of its recovery.

Moreover, even in substantial asset cases where higher stakes increase the probability of a subsequent bankruptcy case absent successful voluntary cooperation, the creditor contemplating action that will place him ahead of the pack often has good reasons to proceed apace. If his recovery comes early enough, before the start of the periods in which preferences and setoffs may be avoided,[106] he will retain an advantage that would otherwise surely be denied him in bankruptcy. When the transfer does occur within the period in which the trustee can avoid a preference, the transferee of property of a consumer debtor will often be protected by simple cost-benefit analysis on the part of the trustee. The trustee may determine that relatively small preferences, common in consumer cases, are not worth the cost of their recovery, and this cost-benefit analysis is formally recognized in the preference statute, which protects transfers of less than $600 by consumer debtors from avoidance as preferences.[107] Of

---

106. Bankruptcy law typically limits the power to avoid preferences and setoff or setoff rights to transactions that occur within 90 days before the filing of the petition initiating the bankruptcy proceeding, but exceptions to the 90-day rule exist in both instances. See 11 U.S.C. § 547(b)(4) (1994); 11 U.S.C.A. § 553 (West 1993 & Supp. 1996).

107. *See* McCoid, *supra* note 59, at 263–65. 11 U.S.C.A. § 547(c)(8) (West Supp. 1996) protects transfers of less than $600 by consumer debtors from avoidance as preferences. Elizabeth Warren and Jay Westbrook summarize the utility of all the trustee's avoiding powers in consumer bankruptcies in one sentence. "[T]he TIB's [trustee in bankruptcy's] avoiding powers are equally available in consumer and business bankruptcies, but their importance is minimal in the former and very great in the latter." ELIZABETH WARREN & JAY WESTBROOK, THE LAW OF DEBTORS AND CREDITORS 177 (1986).

In instances in which the transfer is defeasible under one of the trustee's avoiding powers and the transferee still has a claim against the bankruptcy estate, 11 U.S.C. § 502(d) (1994) assists the trustee in overcoming the cost-of-recovery barrier. That statute mandates disallowance of the creditor's claim if the transfer has not been returned to the

course if bankruptcy does not ensue, the creditor has less at stake in any future contests or negotiations with other creditors. Even the added inducement given creditors entitled to setoff not to exercise their rights prior to bankruptcy is of problematic value. Before it bears cooperative fruit, the creditor must be convinced of the probability of improving or at least not worsening her position. She must believe that the amount of her claim that is not protected by the right to setoff will not increase.

To be sure, there are costs associated with taking a preference or setoff that is later avoided in bankruptcy. The creditor's expenses of obtaining the preference or setoff will come to naught.[108] There is the further danger that the preference or setoff may diminish the debtor's estate in ways that may not be fully rectified by subsequent recovery of the property in the bankruptcy proceeding. Temporary deprivation of capital may compound other financial problems, impeding the debtor's recovery efforts at the expense of all creditors.[109] Loss of the debtor's automobile through levy under a writ of execution may cost the debtor his job, even if the execution lien is later avoided and the automobile recovered by the debtor's trustee in bankruptcy. If the debtor shows promise and the creditor will have a substantial claim remaining after any contemplated recovery, assessing the effect collection may have on the debtor's efforts to regain economic viability may even deter the creditor who has no significant fear of subsequent avoidance of any recovery in bankruptcy.

While considerations such as these may militate against coercive collection in commercial cases, especially those in which substantial claims justify sophisticated monitoring by creditors,[110] they are probably of far less value in curbing coercion in the haphazard world of consumer credit, where monitoring-cost constraints kick in early on.[111] The adage compar-

---

trustee. But this provision will only succeed in producing recovery of a voidable transfer if the prospect of payment on the creditor's remaining claim in bankruptcy exceeds the amount of the voidable transfer, a condition not likely to arise often in consumer cases where little or nothing is available in many instances for pro-rata distribution to general non-priority creditors.

108. *See* McCoid, *supra* note 59, at 265.
109. *See id.*
110. *See id.* at 264–65 (cases in which creditors closely scrutinize the debtor perhaps explain many of the accounts of deterrence of individual collection effort by preference law).
111. While John McCoid seriously questions the efficacy of preference law as a deterrent, *id.* at 264, Thomas Jackson concludes that "showing that it underdeters is not to say that it provides no deterrence at all," JACKSON, *supra* note 2, at 138. Neither writer specifically addresses, however, the additional impediments to deterrence that probably result from decreased monitoring by the holders of the relatively small claims indigenous to consumer credit cases.

ing the value of a bird in hand with two birds in the bush is no more—or less—than a truism in the uncertain world inhabited by collectors of consumer debt.

The foregoing critique of bankruptcy law's inadequacies in furthering the aggregate interest of creditors in consumer cases, fails to distinguish, however, the types of bankruptcy proceedings that consumers may invoke. Creditors' principal criticism of the bankruptcy process—that it results in little or no payments on their claims—may be muted if not silenced altogether in some cases of debt adjustment under Chapter 13 of the Bankruptcy Code,[112] a process that differs dramatically from payment to creditors from liquidation of non-exempt assets under the Bankruptcy Code's Chapter 7.[113] In a Chapter 13 proceeding, the debtor normally does commit some portion of his future earnings to payment of his debts.[114] And bankruptcy's automatic stay[115] protects the debtor from competitive creditor action that might otherwise defeat his payment efforts.

Still, many creditors do not hold Chapter 13 in high regard, and in many instances for good reasons. These reasons are more difficult to ascribe than are the reasons for creditor discontent with Chapter 7 proceedings. Because debtors in Chapter 13 and those in counselor-assisted workouts share a most important characteristic—they make payments to creditors from future income in both proceedings—the results in a successful Chapter 13 case may be similar to those reached in a successful counselor-assisted workout: full payment of creditors' claims. As subsequent comparison of counselor-assisted workouts and Chapter 13 proceedings will show, however, the standards that prevail in practice for repayment to creditors in the two types of proceedings are vastly different, and creditor discontent with Chapter 13 is best understood by contrasting common practices in Chapter 13 proceedings with those practices that garner greater creditor support for counselor-assisted workouts, a subject explored in Chapter Five of this work.

The next section of this chapter examines impediments to the use of workouts that must be effected by the action of the debtor or one creditor

---

112. 11 U.S.C.A. §§ 1301–1330 (West 1993 & Supp. 1996).
113. 11 U.S.C.A. §§ 701–766 (West 1993 & Supp. 1996).
114. *See id.* §§ 1306(a)(2) (augmenting the property of the bankruptcy estate in a Chapter 13 proceeding by "earnings from services performed by the debtor after the commencement of the case but before the case is closed, dismissed or converted" to another chapter of the Bankruptcy Code), 1325(b) (imposing as a condition for confirmation of a Chapter 13 plan that does not propose full payment to objecting unsecured creditors the requirement that all of the debtor's projected disposable income for three years be applied to make payments under the plan).
115. *See* 11 U.S.C.A. § 362 (West 1993 & Supp. 1996).

without the assistance of agencies formed to efficiently marshal creditor support for the undertaking and to monitor the performance of the parties to it. It enlarges upon the previous examination of Kelly's failure to secure the cooperation of her creditors in a workout.

(2) A Return to Kelly's Case: Factors in Addition to Unbridled Creditor Competition that May Defeat a Consumer's Attempt at an Unassisted Workout

The financial debacle in Kelly's case reported in Chapter Two has been presented solely as a product of ruinous competition among creditors, while other factors that could have caused or contributed significantly to it have been conveniently removed by the deft art of assumption. The analysis has proceeded thus far like Mitchell Polinsky's retelling of the tale of three men cast ashore without food or other trappings of civilization on a deserted island. They face the challenge of opening a can of beans washed up by the tide. Two are natural scientists who devise sophisticated means of accomplishing the task without spilling the beans by resort to the principles of their respective disciplines, physics and chemistry. The third, an economist, offers a simpler solution. He proposes that they "[a]ssume a can opener."[116] It is time to admit that factors other than the creditor competition I have assumed may have triggered Kelly's financial demise. There are several.

One of the banks that issued Kelly a credit card may have concluded that she did not need concessions. A bank that grants extensions to a debtor in default, which are not required to effect the debtor's financial recovery, may incur costs without attaining offsetting benefits. Unnecessary extensions given Kelly may have prevented the bank from using the proceeds from prompt recovery of her loan more profitably and thus imposed needless lost-opportunity costs on the bank. The foregone opportunity of the creditor to use the funds derived from prompt repayment in making other loans may not, however, result in a loss of interest income. Nevertheless, in many consumer workouts, a creditor is called upon to lower or forego interest and late charges in addition to extending the term for repayment of principal. Even if the return on principal of the loan in workout equals or exceeds the return the creditor could command from new borrowers, the additional costs and risks of collecting a loan in default will usually cause the lender to rationally prefer an alternative use of its capital.[117]

---

116. A. MITCHELL POLINSKY, AN INTRODUCTION TO LAW AND ECONOMICS 1 (1983).
117. *See generally* PAUL A. SAMUELSON, ECONOMICS 474–75 (10th ed. 1976) (discussing the concept of opportunity costs).

Moreover, unneeded concessions by the bank may have led to more serious consequences. If the bank had alleviated Kelly's pecuniary stress, it may have caused Kelly to discount the debt-management regimen needed to effect her financial metamorphosis, and thus caused her to slip further into the abyss of excessive debt. Whether the debtor's plight results from misfortune, poor debt management or some combination of the two,[118] a first step in achieving a successful workout, where feasible, is early recognition of the problem and the measures, stringent in many cases, that a debtor must undertake to correct it. The importance of early recognition of failing conditions is virtually always addressed, even if only in passing due to its obviousness, in the growing literature on the workout process for business entities.[119] It is no less important in the case of individuals overburdened by consumer debts. Collectors help consumers to the extent that they cause them to face financial problems before debt becomes unmanageable. The use of coercive collection measures in these instances may be benign. This benefit to consumers may result whether the cause of the initial misalignment in debt and income stems from unavoidable interruptions in income[120] or the incurring of non-volitional expenses.[121] Some commentators have a tendency to wax euphoric about the blamelessness of debtors, when all too often these debtors are, to some extent at least, victims of their own poor financial management.

But what if Kelly's need for extensions was, as previously posited, real? She would still face the daunting task of convincing her creditors of her need and, further, of her resolve to use any extensions for enhancing their chances of recovery. Assuming Kelly was able to state her case clearly and forcefully, a counterintuitive assumption in many instances of consumer-debtor advocacy,[122] should that have resolved the matter in favor of the extensions? Collectors who fail to verify a debtor's income, necessary expenses and obligations to other creditors before granting an extension, like those who uncritically rely on the statement that the check is in the mail,

---

118. *Cf.* STANLEY & GIRTH, *supra* note 101, at 47–49 (exploring the causes of personal bankruptcy).

119. *See e.g.*, BUSINESS WORKOUTS MANUAL, *supra* note 99, at 1–7.

120. *See generally* SULLIVAN ET AL., *supra* note 1, at 95–104 (finding income interruptions to be a significant factor in many bankruptcies).

121. *Id.* at 168–69 ("dramatic medical losses are critical for only a tiny fraction of the bankrupt debtors").

122. Elizabeth Hudson's story, recounted in Chapter 2, bears witness to the difficult task faced by even those who are experienced in the art of communication. Ms. Hudson, a former press aide who subsequently authored her own story in a national news periodical, could convince only some of her creditors of the need for extensions on her debts. Elizabeth Hudson, *I Don't Need All the Credit*, NEWSWEEK, July 14, 1986, at 9.

are not infrequently misled and are probably soon advised to seek other avenues of employment.

This is not to suggest that the heroine in our financial tragedy lied; this analysis proceeds mightily on the premise that Kelly and the many like her do not. But that should not prevent this commentator's notice of a fact that collectors notice: that debtors who are under financial pressure are more prone to buy time by overstating the problems of today and the promises of tomorrow than those who aren't. Skepticism is the cachet of successful collectors. The author of one collection manual admonishes his readers to view a debtor's problems in relation to that debtor's past record, character and solvency, and even the state of the economy. Yet he prefaces these remarks with the comment that "toughness" and "tenacity," not "leniency" and "forgiveness," is the "name of the game."[123]

While a creditor can usually verify data supplied by the debtor by resorting to outside sources, the process may prove costly relative to the amount of the debt in consumer cases. Still, some creditors will conference with a debtor, review and verify records and even provide assistance in formulating a budget to accommodate a repayment plan. Some lenders to business entities take on far more onerous duties, in light of the complexities involved in effecting turnarounds of firms, than any that would be required of a creditor assisting in a consumer workout. Moreover, these lenders structure their lending agreements to insure an important role for themselves in advising and influencing the debtor's future decisions. Robert Scott argues that the justification for secured commercial lending—for the blanket security interests often taken by the firm's principal lender to the corresponding disadvantage of the interests of unsecured creditors in the debtor's significantly diminished pool of unencumbered assets—is the leverage the secured party's rights in the collateral provide it in influencing the debtor's business decisions for the benefit of all with an interest in the enterprise.[124]

The creditor that had a security interest in Kelly's principal asset, her automobile, also had greater leverage than her other creditors. But beyond this point, the analogy between commercial and consumer secured lending breaks down. A commercial lender generally has far greater incentives for keeping a firm going than a consumer creditor has in supporting a debtor. The commercial lender may enjoy good rates of return on large loans and the prospect of even more profitable large-scale lending if the firm survives and grows. Without an interest in any substantial ongoing venture, the secured consumer lender is far more apt to resort earlier to seizure and sale of collateral than the commercial lender.

---

123. See KING, *supra* note 6.
124. *See* Scott, *supra* note 35.

Instead of seizing the debtor's property, a creditor may consult and counsel with the debtor and other creditors in an attempt at a workout. Based on the complexity of the issues, a creditor's workout costs will generally be far less in a consumer than in a commercial-lending case. But the amount of the creditor's claim, the factor that imposes a limit on the amount of workout costs the creditor will rationally incur, is usually so much less in consumer cases that it precludes attempts at workouts in many otherwise promising cases.

The views of one writer with experience in the collection of claims from consumers is informative. He favors the use of telephone calls and letters in collection, finding "face-to-face interviews" to be "time consuming." He suggests that the debtor who comes to the collector's office is trying to delay or avoid payment. Therefore, he recommends procedures, such as keeping the debtor standing during the interview, that discourage anything other than immediate payment on the past-due account. While the procedures recommended by this writer may be cost effective in many cases, a collector following his advice would seldom seriously consider assisting a debtor in obtaining extensions from other creditors.[125]

A creditor who has a long-term significant relationship with the debtor and the prospect of a profitable continuing one will more likely bear the costs entailed in arranging a workout than the large-volume bureaucratic creditor, increasingly encountered in consumer cases, who places far less value on a continuing relationship with any one customer. Creditors who lend to low-income consumers regularly assist in workouts by customizing their collection practices to the changing needs of their debtors. But their services entail considerable costs, which are passed on to those debtors.[126]

---

125. *See* JOHN WARREN JOHNSON, CONSUMER CREDIT GRANTOR'S GUIDE TO CREDIT GRANTING, BILLING AND COLLECTING 129–35 (1984).

126. In the 1960's, David Caplovitz studied the practices of merchants marketing consumer durables to low-income consumers and found that these merchants exercised close personal control over payments. They used devices such as weekly, instead of monthly, installment payments and door-to-door, instead of in-store, sales and collections. These credit-sellers became familiar with the debtor through these practices, and the further knowledge they gained by selling to the debtor's friends and relations on the same basis enabled them to properly interpret the reasons for a missed payment and to adjust collection procedures accordingly. As Caplovitz concludes, however, this "traditional" marketing system differs considerably from that of the larger, more formal and bureaucratic system used to market most consumer durables. *See* DAVID CAPLOVITZ, THE POOR PAY MORE 15–30 (2d ed. 1967).

While such a system may meet certain needs in the aberrant low-income market, it exacts too high a price to be seriously entertained as a possible solution to the problem of overly coercive collection practices in the general market for consumer durables. In a study

A further problem is endemic to assistance by a single creditor in verifying debtor need, formulating a workout plan and securing the cooperation of other creditors. These other creditors are necessarily circumspect in their use of data supplied by the assisting creditor due to the inherent conflict of interest generated by that creditor's own stake in the process.

Compassion for the debtor may provide an impetus, in addition to that of financial reward, for the exertion of a creditor's efforts on behalf of a workout. In the initial stages of collection, before the debt has been classified as seriously delinquent, collectors with a proper appreciation of the value of the debtor's patronage may treat her with courtesy and respect for reasons other than selfless compassion. Theodore Beckman and Ronald Foster observe that the collection function is complicated by the need "to build and retain customer satisfaction and good will."[127] These authors apparently reconcile the conflict between retaining customers and collecting delinquent accounts by advocating a collection system that exerts gradually increased pressure on delinquents. They recommend that this system have sufficient flexibility to allow for differences in the financial conditions and personalities of individual customers and the worth of their continued patronage. But they also observe that a creditor cannot achieve the ideal of individual treatment of each account when that creditor has thousands. A collection system that exerts gradually increased pressure on debtors is apparently the best compromise for that practice is a staple of books written for the collector.[128] The report of Kelly's case has cut to

---

conducted shortly after Caplovitz's, the Federal Trade Commission found that while low-income market retailers charged considerably higher prices in these labor-intensive and high-credit-risk transactions, they received a net profit return on net worth considerably lower than their counterparts in the general market. *See* FEDERAL TRADE COMMISSION, ECONOMIC REPORT ON INSTALLMENT CREDIT AND RETAIL SALE PRACTICES IN THE DISTRICT OF COLUMBIA ix-xvi (1968).

Another fact militates against the use of more adaptive collection practices in the general market for consumer credit. The low-income buyer's minimal ability to service debt and his need to establish a personal relationship with a lender in order to qualify for any extension of credit severely limit his universe of potential credit grantors. Therefore the low-income marketeer is far more apt to be the debtor's only installment lender. In comparison, in the general market for consumer credit, the collector must necessarily concern himself with the complexities introduced by the claims of other creditors and the effects of their collection efforts on his own.

127. THEODORE N. BECKMAN & RONALD S. FOSTER, CREDITS AND COLLECTIONS: MANAGEMENT AND THEORY 519 (8th ed. 1969).

128. *See e.g.*, GEORGE O. BANCROFT, A PRACTICAL GUIDE TO CREDIT AND COLLECTION 110–23 (1989); HENRY, *supra* note 6, at 133; JOHNSON, *supra* note 125, at 59–68; KING, *supra* note 6, at 56–60; JON R. LUNN, CONSUMER AND COMMERCIAL COLLECTION DESKBOOK 112–13 (1985); JOHN W. SEDER, CREDIT AND COLLECTIONS 75 (1977).

the chase and omitted much of the subtlety of the art of collection because that subtlety is extraneous to the study of debtors who are located too far out on the collection continuum to warrant their creditors' concern with loss of patronage.

Creditors can use manifestation of concern for the debtor that is feigned to mask selfish motives. A creditor may befriend a debtor who is otherwise surrounded by hostile creditors merely as a ploy in the unrelenting competitive contest, rather than as the initial step in a cooperative undertaking. Debt-collection psychology, as practiced by seasoned collectors, recognizes that the carrot may produce greater individual recovery in some collection contests than the stick. In the early stages of the collection process, the creditor may also attempt to motivate the debtor by appeals to his sympathy, self-respect and sense of fairness. Only if these fail, will the creditor turn to appeals to self-interest, such as maintaining the debtor's credit standing, and fear of unpleasant consequences—the ultimate threat of legal action.[129]

Insofar as the ostensibly friendly collector directs his efforts more toward accelerated recovery of his claim than toward securing those needed equitable concessions from all creditors that further successful rehabilitation, that collector's action is no less destructive of the debtor's objective and creditors' collective interests than unrelenting pressure. It is but an alternative means of applying that pressure, a devious path to the same end sought by the creditor who threatens the debtor with loss of her property. And what better method of currying favor with the debtor could the collector devise than casting himself in the role of a confidant who intercedes on her behalf with other creditors?

There are, of course, creditors who do not try to exact a preference for the leading role they play in a workout. But the temptation to do so does raise suspicions in other creditors that may be difficult to allay. Consolidation loans, in which the assisting creditor pays others with the proceeds of a further advance to the debtor, avoid this conflict-of-interest problem. But there are many instances in which the assisting creditor can not justify assuming all risks of the workout.

Another factor compounds the difficulties inherent in enlisting a sponsoring creditor and securing the cooperation of others in a consumer workout predicated on the assistance of the sponsoring creditor. That creditor must often supplement an initial assessment of the debtor's willingness and ability to effect a turnaround with monitoring during the period in which the workout is being performed. Kelly may have needed continuing help from a creditor to ameliorate the difficult tasks that lay before her.

---

129. *See* BECKMAN & FOSTER, *supra* note 127, at 553–59.

Without that assistance, the risks that she would fail may have reduced the estimated value of any creditor's workout recovery to an amount less than that attributable to that creditor's ability to bring a more effective collection weapon to bear than other creditors. While one of her creditors may have volunteered continuing as well as initial assistance, the prospect of a creditor assuming the transaction costs of the workout must necessarily diminish as those costs mount. Moreover, one can reasonably assume that the non-assisting creditors would perceive the temptation for opportunistic behavior on the part of the assisting creditor to increase over the term of a workout that, like so many do, runs into difficulties. This principle is recognized in bankruptcy where the reach of the trustee's power to avoid preferences is extended from 90 days to one year before the filing of the bankruptcy petition when the creditor is an "insider."[130]

Any one of the factors examined in this section—the creditors' failure to perceive the debtor's need for extensions due to insufficient or unreliable data and the creditors' inability to monitor the debtor's willingness and ability, either initially or during the workout—may have resulted in the failure of Kelly's creditors to cooperate in her rehabilitative efforts. Moreover, two or more of these factors may have pulled in tandem. But one can examine the contributions of these factors to the failure of Kelly's creditors to cooperate from a broader perspective, one which more accurately reflects the role they play in cases like Kelly's. Instead of these various inadequacies and imperfections in information and monitoring, singularly or jointly, being independent causes of failure to achieve cooperative action, they more likely complement a more pervasive cause: each creditor's fear of what other creditors will do. If each of Kelly's creditors was at least somewhat unsure of her need, willingness and ability—and such cases are virtually never free of such nagging doubt—then each must also have recognized that the others harbored similar doubts.

The issue for each creditor then was not simply whether Kelly presented a reasonable case for cooperation in a world in which he was her only creditor, a world virtually nonexistent in this age of numerous credit grantors. Each of Kelly's creditors had to resolve the additional issue of whether Kelly's other creditors could be convinced and stay convinced of the benefits of collective action. As application of game theory in the next chapter will show, the case for any one of Kelly's creditors granting concessions would have considerably less merit if that creditor knew or even suspected that her other creditors would not do likewise. From this perspective, imperfect information and inadequate facilities for monitoring are cast not as independent causes of creditor failure to cooperate but as aspects of a more comprehensive cause—competition among creditors.

---

130. *See* 11 U.S.C. § 547(b)(4)(B) (1994).

A holistic analysis of the causes of creditor failure to cooperate that may result in deleterious creditor behavior comes full circle then back to problems inherent in the race of diligence. Because creditor competition presents the central problem in overly coercive collection practices, it warrants a more detailed study, one using the analytical tools of game-theoretic analysis and particularly Prisoner's Dilemma to ferret out and better understand its causes and cures.

4

Game-Theoretic Analysis of Creditors' Failure to Cooperate

A. Modeling Creditors' Behavior as a Game of Prisoner's Dilemma

As the analysis in the preceding chapter has shown, collection may involve strategies and goals in a contest in which an individual creditor's recovery is determined not by his actions alone but by what other creditors do as well. Because harmful aspects of competition result from the need for each creditor to devise a strategy that is the least vulnerable to the strategies of competing creditors, illuminating insights may be gained by applying concepts from that branch of mathematics studying individuals in mutual interaction—game theory.[1]

The game of Prisoner's Dilemma is particularly useful in analyzing the logical structure of conflicts of interest among creditors. The name, Prisoner's Dilemma, derives from the original anecdote illustrating the game.[2] Two suspected perpetrators of a crime are caught by the police, but there is insufficient evidence to convict either of them of that crime unless one or both testify for the state. Absent such testimony, each can be convicted of only a lesser crime. The suspects are separately jailed, held and questioned by the police. They are given no opportunity to communicate with each other. Each is promised his freedom by the prosecutor if he incriminates the other, provided his accomplice does not likewise incriminate

---

1. *See generally* JOHN VON NEUMANN & OSKAR MORGENSTERN, THE THEORY OF GAMES AND ECONOMIC BEHAVIOR (3d ed. 1953) (a comprehensive analysis of game theory by its formulators).

2. The telling of the Prisoner's Dilemma varies slightly from one source to another. *Compare, e.g.*, ANATOL RAPOPORT & ALBERT M. CHAMMAH, PRISONER'S DILEMMA 24–25 (1965) *with* MARTIN SHUBIK, GAME THEORY IN THE SOCIAL SCIENCES 254 (1982). Both of these authorities attribute the naming of the game to A. W. Tucker, A Two-Person Dilemma (1950) (unpublished paper on file at Stanford University Library).

him. The prosecutor also informs the prisoners of less pleasant possibilities. The prisoner who maintains his silence when his accomplice testifies for the state will receive the stiffest sentence, five years. If both incriminate the other, each will serve a two-year sentence, twice the one-year term imposed on each if both maintain their silence and do not assist the prosecutor. The following matrix reflects these sentences. By convention, the payoffs of the horizontal row player (Prisoner A) precede those of the column player (Prisoner B).[3] The payoffs for the four intersections of the two strategies given the two players may vary within the parameters required by the game, which are subsequently examined.

Prisoner's Dilemma
(numbers in years of prison sentence)

		Prisoner B	
		Incriminate	Maintain Silence
Prisoner A	Incriminate	2,2	0,5
	Maintain Silence	5,0	1,1

Obviously, the collective good of the prisoners, from the standpoint of the minimal aggregate time to be served by both, dictates that each choose not to assist the prosecutor. This also produces an equitable result in that both prisoners serve identical terms of one year. Ironically, the collective good of the players in this prototype of the game of Prisoner's Dilemma is not a social good if we assume that the prisoners are guilty of the crime charged and a longer, not shorter, sentence for each would be just. Theorists engage in a worthier pursuit when the search for a resolution of the player's dilemma occurs in those scenarios in which the players' collective good coincides with the wider social good. Casting the wider social good aside as it plays no part in the issue the game is designed to explore, can what we may assume would be the prisoners' collective goal of spending the minimal aggregate time in jail be achieved by parties unable to communicate and enter into an enforceable agreement?

---

3. *See, e.g.*, MICHAEL TAYLOR, ANARCHY AND COOPERATION 5 (1976).

Each prisoner knows that if he fails to incriminate his accomplice and his accomplice incriminates him, he will receive a five-year sentence. And each prisoner knows that his accomplice knows this, and knows that he knows this. As an examination of the above matrix will reveal, a prisoner is better off incriminating his accomplice regardless of what his accomplice does. As neither prisoner in Prisoner's Dilemma is assumed to be concerned with the welfare of the other or with group welfare,[4] Prisoner A will still choose to incriminate Prisoner B even if he knows B is an irrational deviant who will invariably maintain his silence. Assuming self-interest on the part of the prisoners or the participants in any of the numerous analogues of the game of Prisoner's Dilemma, furthers analysis of those cases in which the goal of cooperation cannot be achieved solely by authority, concern for other individuals, concern for the group or some combination of these factors.

Logically, each prisoner should decide to incriminate the other. There is empirical support for this theoretical solution. Experimental evidence indicates that with anonymous, non-communicating players engaged in single play, mutual incrimination is a reasonably good prediction.[5] The paradox is that by this "logical" behavior each will serve twice the sentence he would have served if both had maintained their silence.

Kelly's case and any of the many like it may be similarly modeled as the "Creditor's Dilemma," although sharing the same surname with the prototype of the game may initially obscure various differences in these siblings, which will require subsequent examination. This adaptation will assume that the creditors of a debtor have a choice of two strategies in collecting their claims. Normally, creditors face not a single choice but a series of decisions in collecting a particular claim. To limit the range of options for each creditor, it is necessary to reduce a series of decisions to the notion of a strategy, which covers a range of choices.[6]

---

4. *See* ROBERT AXELROD, THE EVOLUTION OF COOPERATION 6–7 (1984). The assumption of individual self-interest need not deny, however, any role whatsoever to a player's concern for the group or other individuals. It is only in this broader sense that the concept of self-interest, one that is admittedly laden with troublesome ambiguities, *see, e.g.*, Jeffrey L. Harrison, *Egoism, Altruism and Market Illusions: The Limits of Law and Economics* 33 UCLA L. REV. 1309 (1986), is a necessary element in the game of Prisoner's Dilemma.

5. *See* Gerrit Wolf & Martin Shubik, *Concepts, Theories and Techniques: Solution Concepts and Psychological Motivation in Prisoner's Dilemma Games*, 5 DECISION SCI. 153 (1974).

6. *See* R. DUNCAN LUCE & HOWARD RAIFFA, GAMES AND DECISIONS 6–10 (1957).

The first strategy, "cooperation," consists of adjusting the overextended debtor's obligation, usually by extending the term for amortization of the debt. But when necessary, a creditor can supplement its acceptance of reduced payments or its grant of a temporary moratorium on all payment by forgoing late charges, reducing interest rates or forgiving payment of part of the principal amount of the obligation if the workout is successful.

The second strategy, "coercion," consists of rigid insistence on immediate payment of all amounts in arrears, which may include installments that are due only by the creditor's application of an acceleration clause in the contract that causes the entire obligation to become due upon default on any installment. The creditor employing a coercive strategy often endeavors to promote his claim over others by the use or threat of using more punitive measures than his fellow creditors.

The reader familiar with the literature of Prisoner's Dilemma will note a difference in the names assigned the strategies in Creditor's Dilemma and those conventionally used in general discussions of Prisoner's Dilemma. Customarily, players' strategies are described for their effect on the other player—"cooperation" or "defection." Departure from that convention stems from the immediate focus of this study on the effect creditors' actions have on the debtor's rehabilitation efforts. From this perspective, the two choices are titled "cooperation" and "coercion." The parallels with conventional designations, however, should be readily apparent. Cooperation with the debtor, which enhances the chances of recovery by other creditors in cases in which the model of a Prisoner's Dilemma applies, equates with the conventional strategy of cooperation. And a strategy of coercion with respect to the debtor in an instance in which the aggregate recovery of creditors is reduced, is tantamount to the conventional strategy of defection from the perspective of other creditors.[7]

But a coercive strategy does not preclude manifestation of concern for the debtor by the creditor employing it, provided his actions are still designed to establish his priority in payment—in this instance by feigned friendship instead of fear—and he still denies needed concessions. Nor does a cooperative strategy preclude the use of uncompromising efforts to enlist and maintain the debtor's commitment to a viable workout plan in which all creditors are treated equitably. Yet the critical distinction between these two options justifies the designations chosen. One, cooperation, signifies flexibility and fit to the debtor's and creditors' collective needs; the other, coercion, action detrimental to those needs.

Quantifying the assumptions previously made in cases like Kelly's could produce the following matrix. For simplicity, assume only two creditors,

---

7. *See, e.g.*, AXELROD, *supra* note 4 at 7–9.

with claims against their common debtor of $6,000 each and with equal powers to coerce payment. Again, the payoffs to A precede those to B.

Creditor's Dilemma
(numbers in thousands of dollars)

		Creditor B	
		Coerce	Cooperate
Creditor A	Coerce	2,2	5,1
	Cooperate	1,5	4,4

If both creditors pressure the debtor by using coercive strategies, she soon reaches the point where, like Kelly, she abandons her workout efforts, and each creditor collects only one-third of his $6,000 claim. The highest aggregate recovery is achieved when both creditors cooperate with the debtor's workout efforts and each recovers two-thirds of his claim, $4,000. This is the second best payoff for each creditor. Either may achieve a slightly better one only at a substantial cost to the other and the aggregate recovery of both creditors, the payoffs resulting from each creditor's use of coercion against the other's use of the cooperative strategy.

As in Prisoner's Dilemma, a creditor with no means of entering into a binding agreement with others will do better using the noncooperative strategy regardless of what his opponent does. If B cooperates, A's use of coercion garners him $5,000 instead of the $4,000 he would have received had he cooperated. And if B employs coercion, A's use of the same strategy produces $2,000 compared with only a $1,000 recovery for him had he cooperated. The same incentive to employ the coercive strategy obtains when the contest is viewed from B's perspective. Recognizing the dominance of the coercive strategy, each creditor will employ it and receive only $2,000, one-half of the $4,000 awarded each for mutual cooperation. The dilemma is that, while each obviously prefers $4,000 to $2,000, neither can afford to cooperate unless he knows in advance that the other will do likewise. "Prisoner's Dilemma is simply an abstract formulation of some very common and very interesting situations in which what is best for each person individually leads to mutual defection, whereas everyone would have been better off with mutual cooperation."[8]

---

8. AXELROD, *supra* note 4, at 9.

While there are striking parallels in the Creditor's and Prisoner's Dilemmas, the former fails to replicate the latter in certain obvious respects. These distinctions require further analysis.

For ease in exposition, the original metaphor incorporates a simple means of providing the players with needed information. In the prototype, each prisoner (1) knows the payoffs, (2) knows that the other player knows them and (3) knows that the other player knows that he knows them. An omniscient prosecutor reveals these matters at the time he presents the options to the prisoners.

In Creditor's Dilemma, neither certainty nor dissemination of information is quite so neat. No one occupies the prosecutor's role to inform the players of the payoffs. Of even greater import is the threshold difficulty of reducing the payoffs to certainties in the game creditors play. Values based on assessments of the debtor's future performance are difficult to ascertain with any degree of precision, and it is unlikely that creditors will assess them identically.

Still, one may conclude that in some ordinary collection cases, in which there is no agent to knowledgeably establish payoff parameters and inform the contestants of those parameters, creditors possess sufficient data to validate use of the techniques of Prisoner's Dilemma to predict their actions. To evaluate this conclusion, one must ask what requirements the payoffs contained in the 2 X 2 matrix must meet for the creditor's contest to qualify as one of Prisoner's Dilemma?

The answer is that the requirements posited by game theorists are not nearly so exacting as presentation of precise data in matrices for the game would suggest. In lieu of requiring that all players assign absolute values to payoffs, Prisoner's Dilemma mandates compliance with two rules of inequality.

In the order of outcomes, the highest reward for a creditor must occur when he chooses coercion in an instance in which the other creditor cooperates. This temptation reward (T) is paired with the lowest outcome, the sucker's payoff (S), thrust upon the cooperating player on the same play. Among the two intermediate payoffs, both of which result from choices of the same strategy by each creditor, the reward for mutual cooperation (R) must exceed the punishment for mutual coercion (P). Hence, the first rule of inequality is $T > R > P > S$.[9] Based on this rule, a creditor contemplating S will be motivated to use a coercive strategy to get at least P, while one contemplating R will be attracted to the coercive strategy to get T.

A model based on Prisoner's Dilemma must also meet a second rule of inequality. If an even chance of receiving temptation (T) or sucker (S) pay-

---

9. See id. at 9–10; RAPOPORT & CHAMMAH, supra note 2, at 33–34.

offs, of exploiting or being exploited, produced a higher payoff than R, the reward for mutual cooperation, then the players could escape their dilemmas by relying on chance to divide the greater gains from taking turns exploiting one another in repeated contests. And creditors do engage each other in iterated games. To foreclose this resolution of the dilemma, the reward for mutual cooperation must be greater than the arithmetic mean of the sum of the temptation and sucker payoffs, or $R > (1/2)(T + S)$.[10]

Sufficient conditions exist for Creditor's Dilemma when each player assesses the value of the four payoffs in accordance with these two rules of inequality. But increasing one payoff while keeping others constant may affect the choices of actual as opposed to purely rational players, even though the revised payoffs still comport with the rules of inequality. In an experiment in which 70 pairs of college students played Prisoner's Dilemma games 300 times in succession, researchers reported that cooperative responses between players tended to increase with relative increases in R or S and decrease with relative increases in T or P.[11]

While it simplifies the presentation of Creditor's Dilemma to assume that the players assign identical values to each of the four payoffs, this symmetry is not required.[12] Moreover, neither player need assign an absolute value to any of the payoffs, provided his rank ordering of all and relative weighing of the relevant three comply, respectively, with the first and second rules of inequality. Finally, it is not necessary that the players measure the value of their payoffs by the same determinants. One commentator suggests that a cooperating bureaucrat who leaks a story to a journalist might get rewarded by a chance to have a policy argument presented in a favorable light, while the cooperating journalist might get rewarded with another inside story.[13]

While recovery of their respective claims provides creditors a common determinant of value, one creditor may supplement his payoff from recovery of his claim with other values, which his opponent may share not at all or only nominally. When this occurs, these incremental values will be reflected in one creditor's perception of the payoff matrix, though not in the other's. For example, one creditor may be concerned with the effect of collection practices on the subsequent patronage of a rehabilitated debtor and those friends and relatives that the debtor influences. This creditor's

---

10. *See* AXELROD, *supra* note 4, at 10; RAPOPORT & CHAMMAH, *supra* note 2, at 34–35.

11. *See* RAPOPORT & CHAMMAH, *supra*, note 2, at 33–49. For other effects of iterated games on players' choices of strategies, *see* Chapter 5.

12. *See* AXELROD, *supra* note 4, at 17.

13. *Id.*

additional increment of value may benefit more from the employment of one strategy—normally cooperation—than the other. Conversely, one creditor, but not the other, may assign additional utility to the coercive strategy for its perceived enhancement of his reputation as a forceful collector with other delinquent debtors. Of course, any creditor's projection of payoffs from all sources must still meet the tests of the two rules of inequality.

These rules pertaining to priorities of outcomes and their relative values are but a subset of a larger assumption of game theory—that players strive to maximize utility. But one player's utility may not conform with others' widely shared notions of the goal or goals of the contest and, thus, with the payoff matrix as commonly perceived. In some instances, that player's aberrant values may not merely supplement those commonly shared by others but may replace them entirely. Duncan Luce and Howard Raiffa furnish an example of how these idiosyncratic values may assume paramount, even sole importance.[14] When poker is played for money, a player should choose strategies based on the money outcome, but players who enjoy bluffing for its own sake may do so with no regard for the money outcome. Likewise, a creditor may wish to punish a debtor for what he perceives as that debtor's profligate ways, regardless of the effect that punishment may have on the recovery of his claim. While such distortions of commonly shared goals diminish the explanatory power of a creditor's contest modeled on Prisoner's Dilemma, their incidence is probably rare. For a large, bureaucratic creditor, the bottom line on collection practices is the bottom line on the financial statement.

Simplicity in presentation alone provides sufficient reason for limiting each creditor's assessment of the values in the preceding matrix to recovery of his outstanding claim. But the exclusion of other considerations—and the two picked as illustrative would appear to be the most significant—may be justified on other grounds as well.

The contribution of coercive practices in a single case to the creditor's reputation for forcefulness, a peripheral concern, must pale when compared with the more tangible goal of maximizing actual recovery in that case. A case that otherwise clearly meets the tests of a Creditor's Dilemma is unlikely to be disqualified solely by the value a creditor attaches to a coercive strategy in that case furthering his reputation for diligent collection efforts. Concern with that reputation, however, may influence the creditor's conduct in other instances. Where mutual cooperation by creditors offers no clear promise of increased recovery and a creditor is contemplating the use of more stringent measures to increase pressure on the debtor, the cred-

---

14. *See* LUCE & RAIFFA, *supra* note 6, at 3–6.

itor's decision may be influenced by the need to convince other delinquent debtors of his persistence in collecting claims. In some cases, this complementary factor may cause a creditor to invest more resources in coercive effort than is justified by probable recovery in the case.

The second consideration of a creditor other than maximizing recovery of his claim is also not likely to influence that creditor's choice of strategies in many instances. In today's mass markets for retail finance, where creditors carry literally thousands of consumer accounts and employ standardized procedures, which permit little that is cost effective in the way of individual attention to particular accounts, it is difficult to conceive of any creditor ascribing any significant value to the prospect of subsequent transactions with an individual who has already commanded the attention of his collection department. In a less mobile society with more parochial lenders, loss of goodwill of a consumer, her extended family and friends may have carried appreciable weight. Today, however, such considerations would appear to be limited primarily to some extenders of commercial credit who do a significant amount of their business with a few customers.

The conditions that result in a commercial supplier's dependency on one large buyer may stem from institutional economies that also make the buyer dependent on the supplier. Oliver Williamson finds that these conditions exist where highly specialized inputs of production preclude other ready sources of supply but the buyer's needs are insufficient to result in equal economies of scale through self-production.[15] In consumer transactions, where there is no common parallel to this symbiosis, the parties will make far less effort to sustain the business relationship. Even if the buyer has frequently recurring needs, the value of his patronage is generally no greater than that of numerous other customers. And the value of any consumer's patronage must be further diminished when, as is often the case in recent years, the credit grantor is a card-issuing bank that does not derive a profit or other income, other than the small transaction fee imposed on the merchant seller, from the sale of the goods or services.

Another factor contributes to a creditor assigning little value in any one collection case to his concern with either the goodwill of customers or his reputation as a diligent collector. Obviously, any value assigned to one of the goals must be reduced by the value the creditor assigns the other, for the actions that further one are the antitheses of those that further the other. A creditor's need to balance customer goodwill with a reputation for collection that militates against his claim being ignored or subordinated to those of other creditors is reflected in collection systems that employ grad-

---

15. *See* Oliver E. Williamson, *Transaction Cost Economics: The Governance of Contractual Relations*, 22 J. L. & ECON. 233, 250 (1979).

ually increased pressure on delinquent debtors. When more stringent collection measures seem necessary, the use of a collection agency may afford the creditor some protection against loss of customer goodwill. Some customers, even some delinquent debtors who are the objects of the collection agency's harsh actions, fail to hold the creditor responsible for the actions of the agent he employs.

One or both of the players in a game of Creditor's Dilemma may also include non-pecuniary considerations in assessing the values of the various payoffs. But instances in which the debtor is a close friend or relative of a creditor are also too isolated to warrant serious attention in a world of institutional grantors of consumer credit.

Any attempt at an exhaustive compendium and valuation of all factors that may influence any given creditor's perception of the payoff matrix is doomed to failure for the same reason that an attempt at a complete decision theory of human behavior would be absurd. Russell Hardin contends that "[o]nly in an assumed context can one sensibly be asked whether one's action was rational."[16] He offers two objections to expanding the calculus of an individual's decision-making process in regard to participation in collective action beyond the commonly recognized goals of others in his group. First, the additional factors, which could only be crudely measured as they stem from a host of motivations, would not be worth measuring, since most of the relevant behavior may be explained by the narrowest assessment of costs and benefits. Second, the expanded variables may only explain the conduct of certain members of the group, not that of most of its members.

The appearance of absolute certainty in the payoffs contained in the matrix for Creditor's Dilemma must be qualified for a further reason. No outcomes reflect payment of the entire $6,000 claim of either creditor, although there are doubtlessly numerous cases in which creditors do recover full payment, including interest for extensions, from debtors in distress. All payoffs in the matrix are discounted, however, to reflect that no intersection of strategies guarantees complete recovery for either player in any particular case.[17] They recognize only that some combinations hold greater promise than others. Again, the payoffs contained in the matrix for Creditor's Dilemma are meant to suggest, not to be, the perceptions of either creditor in the game.

---

16. RUSSELL HARDIN, COLLECTIVE ACTION 14 (1982).

17. *See e.g.*, *The Business in Trouble—A Workout Without Bankruptcy*, 39 BUS. LAW. 1041 (1984) [hereinafter *A Workout Without Bankruptcy*] (edited transcript of a panel discussion presented on August 2, 1983, by the Commercial Financial Services Committee of the then Section on Corporation, Banking and Business Law of the American Bar Association).

In Prisoner's Dilemma, there is no doubt that mutual cooperation produces the best collective outcome for the prisoners. The prosecutor, assumed to be omnipotent in establishing the outcomes of the game, assures that result. In Creditor's Dilemma, mutual cooperation produces only a probability of the best collective outcome for the creditors. As even the most promising of workouts may fail, creditors ordinarily insist that each creditor bear the risks of failure equitably, that the extensions of payment and any other concessions of one are commensurate with those of the other. Reaching an agreement that equitably shares the risks of the workout imposes an additional dimension and further transaction costs on a cooperative solution to a Creditor's Dilemma. In consumer workouts, creditors frequently agree to pro-rata extensions of all or substantially all unsecured claims. In commercial cases, where workout may require not only extensions of existing debt but further extensions of credit and resolutions of difficult business decisions, negotiations among creditors may be highly complex.

Contests among consumers' creditors frequently arise in cases where there is no impartial agent to credibly ascertain and disseminate information on outcomes. But the tests imposed by the rules of inequality are not so stringent as to preclude analyzing some such contests as games of Prisoner's Dilemma. While the obstacles to modeling a collection contest on the prototype of the game may be surmounted by applying the flexible rules of inequality to facilitate creditors' determination of payoff parameters in accordance with the requisites of Prisoner's Dilemma, the ability of creditors to do so requires further elaboration.

Implicit in the concept of any game is the players' awareness of the contest and its rules. Before each creditor reasons as Prisoner's Dilemma predicts, she must know that she is engaged in a contest with another creditor and that both view the payoff parameters in accordance with the rules of inequality. She must also know that the other creditor knows that she knows this. But these barriers too are not necessarily insurmountable in cases in which the services of a facilitating agent are absent.

A creditor may know she is engaged in a contest if a delinquent debtor tells her that the collection line has already formed. But if out of embarrassment or otherwise the debtor fails to do so, it is difficult to conceive of a creditor so myopic as to believe that she alone extended the debtor credit. And if one creditor equates the outcomes of her collection strategies with the rules of inequality, certainly there are some cases in which that creditor believes that others will do likewise.

In any case in which the preceding string of assumptions holds, creditors will perceive their collection efforts as a game of Prisoner's Dilemma. This does not mean that collectors familiarize themselves with works on that subject and expend time and effort constructing matrices or otherwise formally engaging the rules of game theory. Riding herd on those who stray from their financial obligations affords little opportunity for these harried outriders of a credit economy to pursue such abstractions. The utility of Prisoner's Dilemma is not that it furnishes its players with a guide for their actions, but that it furnishes a general framework for analyzing those actions. Characterizing a collection contest as a game of Prisoner's Dilemma simply means that the creditors face the same frustrations in choosing their strategies as do the prisoners in the prototype of the game. And creditors can readily comprehend these frustrations, even if they are not familiar with the concept of Prisoner's Dilemma.

The issue in Kelly's case was why her two principal unsecured lenders, the credit-card issuing banks, continued their efforts to maximize short-term recovery when their mutual cooperation in the workout she proposed held greater promise. And as the initial analysis of the case assumed that both banks shared this view that their aggregate recovery would be enhanced by cooperating in a workout, Kelly's case was first cast as a game of Prisoner's Dilemma. The explanation offered earlier—that each bank feared the other would take advantage of its cooperative stance by employing a coercive one—did not employ nor need to employ the language or formal rules of game theory. The coercive action of each bank was attributable to their schooling in the practical aspects of creditors' and debtors' behavior and not the abstract theory of Prisoner's Dilemma. Each knew that debtors respond to pressure, that they are likely to pay more insistent creditors first. And as each creditor knew that the other creditor shared this abiding tenet of his faith, the race of coercive diligence was on.

B. The Transaction-Costs Barrier to a Cooperative Solution of the Creditor's Dilemma

(1) The Scope of the Barrier

Before concluding, however, that competition necessarily results in an equilibrium of mutual coercion, a significant final step in the analysis remains. Motivation for a cooperative solution to the dilemma is a corollary of the same forces in the matrix that frustrate choice of that cooperative solution. The players in both the Prisoner's and Creditor's Dilemma are provided an incentive, even though acting on purely selfish grounds, to cooperate with each other: the prisoners by maintaining their silence and the creditors by extending their terms of payment. Yet, the prisoners will not likely cooperate. Stone walls, barred windows and guards enforce the prosecutor's decision to separate the prisoners and prevent any communication that may result in either an enforceable agreement between them or, a not unlikely alternative in a prison setting, credible mutual threats. Again, the rococo embellishments of the prototype parody life's more commonly encountered analogues.

The impediments to a cooperative solution to a Creditor's Dilemma, suggested earlier by Kelly's case, were transaction costs. Some cost-effective means must be devised to ensure that no creditor succumbs to the lure of the temptation payoff and pairs another creditor's cooperative strategy with his coercive one. As vivid symbols of constraint, these transaction costs pale in comparison with the devices so readily available to the prosecutor. Yet the bonds they place upon actors on the economic stage may be as constraining as cells to prisoners.

Transaction costs play a significant role in the analysis of both economics[18] and law.[19] To an economist explaining their effect on the market exchange of commodities, they include those costs "that stem specifically from the contending wills and property interests of the parties,"[20] such as expenditures incurred in the communication of offers, comparisons of alternatives, negotiation of contracts and verification of the performance of those contracts.[21] Commentators on the basic principles of law and eco-

---

18. *See e.g.,* JACK HIRSHLEIFER, PRICE THEORY AND APPLICATIONS 386–421 (4th ed. 1988) (an analysis of transaction costs and how they affect market equilibrium).

19. Judge Posner observes that transaction costs "are a recurrent theme" in his comprehensive study of law and economics, RICHARD A. POSNER, ECONOMIC ANALYSIS OF LAW 367 (3rd ed. 1986), a statement that is amply confirmed by numerous references to the subject in the index of his book, *see id.* at 665.

Credit for seminal studies on the importance of transaction costs in the analysis of law goes to Ronald H. Coase, *see* Coase, *The Problem of Social Cost,* 3 J. LAW & ECON 1 (1960); Coase, *The Nature of the Firm,* 4 ECONOMICA 386 (n.s. 1937), *reprinted in* READINGS IN PRICE THEORY 331 (George J. Stigler & Kenneth E. Boulding eds. 1952). In 1991, Coase was awarded the Nobel Memorial Prize in Economic Science for his work in transaction costs. *Friction Theorist Wins Economics Nobel,* WALL ST. J., Oct. 16, 1991, at B1. Speaking at a meeting of law-school teachers earlier that year, Coase remarked that "the key concept in all my analytical schemes is transaction costs." Address by Ronald Coase, Luncheon of the Section on Law and Economics of the Association of American Law Schools (January 4, 1991) (unpublished).

Other significant contributions to the study of transaction costs include OLIVER E. WILLIAMSON, MARKETS AND HIERARCHIES: ANALYSIS AND ANTITRUST IMPLICATIONS (1975); Williamson, *supra* note 15, and Guido Calabresi, *Transaction Costs, Resource Allocation and Liability Rules—A Comment,* 11 J. LAW & ECON. 67 (1968).

20. JACK HIRSHLEIFER, PRICE THEORY AND APPLICATIONS 231 (2d ed. 1980).

21. *See id.*

nomics also give the concept wide scope. Judge Posner defines transaction costs as "the costs involved in ordering economic activity through voluntary exchange,"[22] while Mitchell Polinsky subsumes within the term "the costs of identifying the parties with whom one has to bargain, the costs of getting together with them, the costs of the bargaining process itself, and the costs of enforcing any bargain reached."[23]

The very breadth of the concept of transaction costs, however, has engendered criticism of the device as an analytical tool. One commentator complained that "[t]ransaction costs have a well-deserved bad name as a theoretical device... [partly] because there is a suspicion that almost anything can be rationalized by invoking suitably specified transaction costs."[24] While the complaint doubtlessly has merit, transaction costs broadly defined appear to be the sole barrier to obtaining rational cooperation by creditors in workouts that appear promising.

In recognizing transaction costs as the culprit that may defeat mutually beneficial exchanges of coercive for cooperative strategies among creditors,[25] my intent is again to give the concept wide meaning, to include within its parameters all costs associated with obtaining concerted creditor action by agreement, tacit or formal.

For any creditor who will consider undertaking the task of sponsoring a workout, initial transaction costs include those of obtaining and verifying relevant data concerning the debtor's income and necessary living expenses and the extent and terms of her obligations. After concluding that the debtor has a cash-flow problem that transcends mere reasonable curtailment of discretionary expenses, the creditor must consider whether workout is feasible and, if so, he must assess the probability that the debtor will use any extensions given to promote her financial recovery, rather than use the relief from creditor pressure to slip further into debt. Consideration of further data that reflect on the debtor's financial character, such as her past credit history, stability of employment and the reasons for her overextension, may assist him in making this judgment, but in many cases the call will still be a difficult one.

---

22. POSNER, *supra* note 19, at 231.

23. A. MITCHELL POLINSKY, AN INTRODUCTION TO LAW AND ECONOMICS 12 (1983).

24. Williamson, *supra* note 15, at 233 (*quoting* Fischer, *Long-Term Contracting, Sticky Prices, and Monetary Policy: Comment*, 3 J. MONETARY ECON. 317, 322 n. 5 (1977)).

25. *See* William C. Whitford, *A Critique of the Consumer Credit Collection System*, 1979 WIS L. REV. 1047, 1077 (transaction costs frequently preclude resolving conflicts between single creditors and creditors as a class by bargaining and agreement among them).

Where the sponsoring creditor bases his decision in favor of the merits of an extension plan on a less than clear probability that it will enhance aggregate creditor recovery, the costs of the additional steps he must take to bring the workout to fruition will increase. The sponsoring creditor must contact the other creditors, or certainly the principal ones, and convince them of the benefits of the plan. But the absence of precise criteria for measuring the workout's chance of success will increase the costs of securing the other creditors' cooperation in two ways. Mutually beneficial exchanges may be defeated simply by the costs of negotiations that are necessary to allocate the benefit they promise among the parties, and this impediment must assume increasing importance where the benefit is questionable and difficult to measure. Moreover, mutually beneficial exchanges may be defeated by the need to resolve the more fundamental issue of whether the proposed transaction offers any benefit for the parties to share.

Obviously, a creditor's estimate of the amount of his recovery in a contest with other creditors, the creditor's alternative to entering into any proposed workout agreement, also poses a problem of valuation. But there is a presumption, presumably empirically derived, that a creditor's recovery is generally increased by application of continuing pressure for immediate payment. In the words of one commentator, "[e]xperience has shown that there is a definite correlation between the length of time debts are unpaid and the volume of resulting bad debt losses."[26] As analysis in the next subsection will show, the burden of proof in a collection case is customarily assigned to the proponent of the extension plan.

An additional element in the transaction costs of securing the cooperation of creditors in workouts is not inconsequential. Someone must continually and impartially monitor the actions of the debtor and all her creditors during the implementation of the plan to ensure their good-faith compliance with its terms. Even when the sponsoring creditor is willing to undertake this additional task, an all too evident conflict of interest seriously impairs his ability to function as an impartial agent.

Obviously, transaction costs in all their facets will defeat a mutually beneficial exchange only when those costs exceed the perceived value of the exchange to the parties.[27] There are reasons, however, for believing that transaction costs impose more significant barriers on workouts than on typical market transactions, such as ordinary sales of goods or services. Judge Posner attributes high transaction costs to two primary factors: (1) a large number of parties to a transaction and (2) the inability of the par-

---

26. ROBERT H. COLE, CONSUMER AND COMMERCIAL CREDIT MANAGEMENT 371 (5th ed. 1976).

27. *See* POSNER, *supra* note 19, at 55.

ties to the transaction to deal with others, a condition economists term a bilateral monopoly.[28] He observes that "costs of transacting are highest where elements of bilateral monopoly coincide with a large number of parties to the transaction—a quite possible conjunction."[29] Both elements are normally present in workouts.

Although I have cast, for ease in exposition, only two players in my rendition of the Creditor's Dilemma, the number of parties in a workout will frequently exceed the two ordinarily found in a sale of goods or services. And the certain presence in workouts of a bilateral monopoly, an atypical condition in other market transactions, may also unduly protract the process of negotiating. Creditors of a debtor are unable to deal with anyone other than each other. This gives each creditor an incentive to hold out for the promise of a more expedited payment on his claim than that made to other creditors as the price of any concession on his part. The creditor who holds out for preferential treatment need not appear unreasonable. He may base his claim on his real or feigned belief that the plan offers less chance of benefit than that perceived by its proponents.

The context of a workout adds a further element that magnifies transaction costs. The parties must negotiate in a setting of financial distress. This setting makes even more acute the conflicts of interest and incentives for deception that are present in the transaction.[30]

While transaction costs may impose insurmountable barriers on creditors obtaining the benefits of a workout, do those that were present in Kelly's case appear that formidable? So long as the assumption holds that each of the two principal creditors from whom she sought extensions believed that mutual concessions would further both of their individual recoveries, the answer would appear to be no. In our exchange economy, the perceived mutual advantage of traders overcomes the costs of the exchange process in literally millions of transactions daily. In these exchanges, transaction costs reduce the aggregate benefit of traders and are thus a form of market imperfection. Traders, however, often willingly incur these costs to reduce other imperfections that would impose even more significant defects on the exchange process.[31] A seller of property is willing to incur a brokerage fee, such as that paid to a real estate agent or stock broker, in any instance in which incurring that agency cost appears preferable to doing business in a market even more seriously flawed by severely

---

28. *Id.* at 54.

29. *Id.* at 55.

30. *See* Stewart C. Myers, *Determinants of Corporate Borrowing*, 5 J. FIN. ECON. 147, 158–59 (1977).

31. *See* HIRSHLEIFER, *supra* note 20, at 403–8.

limited access to potential buyers.[32] Although the benefits of the workout in Kelly's case were of a lesser magnitude than those posited in the matrix for Creditor's Dilemma, as smaller claims were at stake, it is nevertheless difficult to comprehend how the transaction costs of either of Kelly's creditors obtaining the cooperation of the other could have exceeded the probable benefits of such action. A sufficient measure of the force that triggers voluntary undertakings should have been present even if the tradeoffs promised Kelly's creditors only the probability of small net gains.[33]

Kelly's case presents a collection setting more appealing than most for predicting her creditors' assumption of the transaction costs of a workout. The coercive practices that resulted in Kelly's abandonment of her repayment efforts resulted from the actions of only two of her creditors, her credit-card issuing banks. As previously noted in the report of her case, Kelly was also pursued by a utility company that threatened discontinuance of essential services and a secured lender that threatened repossession of the automobile that she needed to get to and from work. While Kelly's rehabilitation plan gave these creditors a preference in payment of arrearages over the claims of the banks, it is customary in workouts to protect assets that are vital to the success of the debtor's efforts. This practice is followed in counselor-assisted workouts where equitable treatment of creditors is otherwise the norm. And while Kelly's problems with these banks were exacerbated by the fact that they were located in another state and in a city some 300 miles from her residence at the time she attempted to obtain their cooperation, this fact does not appear to be unduly significant in this age of increased mobility.

Kelly's failure to obtain the cooperation of the banks may have resulted from one or more of several factors, all of which may be subsumed, however, under the rubric of transaction costs. Even though each bank only had to secure the cooperation of the other, and not that of numerous other creditors, they may not have been able to agree on how to share the risks and benefits of the workout. One or both may have insisted upon receiving a greater than pro-rata share of Kelly's payments until its claim was paid. And even if an agreement could have been reached, each bank may have feared that the collectors of the other would have subsequently ignored it. It seems more likely though that Kelly's case released the complete array of transaction costs that may defeat beneficial action by creditors, and therein lies its value as the paradigm of this study.

---

32. *See id.* at 405.
33. *Cf.* ARTHUR ALLEN LEFF, SWINDLING AND SELLING 118–30 (1976) (the most common ploy in selling, "The Squaresville Pitch," is to credibly explain the seller's marginal cost advantage and his reason for sharing that advantage with the buyer).

Abandoning the hypothesis that both banks could discern the benefits of the plan simply from Kelly's presentation of it unleashes a difficult threshold issue. Did the banks believe that they could improve their prospects for payment if both accepted Kelly's proposal? A negative answer on the part of both banks may have ended the matter. But it is not unlikely that Kelly's self-advocacy may have convinced one or both of the banks that there was enough merit in her proposal to overcome the inertia of their coercive practices and justify giving her a chance at workout if either of them were her only creditor, but insufficient merit to do what Kelly required in a world requiring concessions from two creditors. Neither bank was willing to advocate acceptance of Kelly's proposal, with or without additional investigation of the facts, to the other. Nor was either bank willing to grant a unilateral extension. The use of a liberal extension by one bank, even without any extension by the other, may have provided Kelly with a means to support her workout efforts. Creditors frequently give extensions to debtors who have suffered reversals without regard to the actions of other creditors, provided the debtor's plight is not so extreme as to raise justifiable fears that he will fail. Ironically, so far as the debtor is concerned, only when extensions are most needed do they appear to be conditioned on the additional, often difficult burden of obtaining concerted creditor action.

Before examining in Chapter Five how all transaction costs of resolving a Creditor's Dilemma may best be systematically contained within the severe cost constraints that characterize the collection of consumer debt, the next subsection further explores the effects of a creditor's failure to properly discern the nature of the contest. Inadequate or incorrect data, judgmental error, or both may distort a creditor's perception of the setting that dictates her choice of collection strategies.[34] Altering that creditor's erroneous perception to comport with reality imposes further transaction costs, which must also be readily contained in any cost-effective resolution of a Creditor's Dilemma. This work introduced the causes of this misperception problem in the earlier analysis of Kelly's case by relaxing the initial assumption that her creditors discerned the benefits of their mutual cooperation. Following a similar relaxation of that assumption in this subsection, the problem has been further explored in the context of examining the elements of transaction costs. What remains to be done is to explore, with further assists from game-theoretic analysis, how creditors' misperceptions of the collection setting lead almost invariably to their choice of coercive over cooperative strategies. The following section will address

---

34. *Cf.* Charles J. Goetz & Robert E. Scott, *Principles of Relational Contracts*, 67 VA. L. REV. 1089, 1115 (1981) (information costs may render ineffective the enforcement of a best-efforts obligation by the standard legal method of suit on the contract).

how the strategies dictated by those misperceptions foreclose resolution of the Creditor's Dilemma by misguided, chance applications of cooperative strategies.

(2) A Closer Look at Creditors' Misperceptions of the Collection Setting: Zero-Sum Games and No Contests

While players of Prisoner's Dilemma are in conflict with each other, there is an important difference in the nature of their relationship and the relationship of the players in a significantly different contest that creditors frequently play—the zero-sum game. In the zero-sum game, the winnings of one player are the losses of another, so that the algebraic sum of the payoffs to each player always equals zero.[35] Players of zero-sum games face no dilemma in formulating a strategy. As a pure conflict of interest exists, each player unambiguously acts in her self-interest by picking the strategy designed to maximize her own payoff. There is no role for cooperation to play as the players cannot increase their aggregate recovery. Examination of the dynamics of a zero-sum game will not only illustrate the futility of cooperation by creditors in such a contest but will also highlight the unique role the incentive for cooperation may play in resolving a creditors' contest that is a game of Prisoner's Dilemma. While the pure conflict of interest present in a zero-sum game is of little psychological interest, the partly coincident and partly opposed interests of the players in Prisoner's Dilemma does provide a study in internal conflict, which captures the imagination of students of psychology and serious literature.[36]

The following matrix presents a zero-sum contest between two creditors. The matrix contains payoffs to the horizontal row player (A) from the column player (B). Since the game is zero-sum, the payoffs to B from A are the same numbers listed in the matrix with the opposite sign attached. A receives payoffs greater than zero, and B receives those less than zero. Obviously, the debtor makes the payments to both creditors, and neither creditor pays the other. To present the game, however, as one between rival creditors from a common debtor, the payoffs, although made by the debtor, represent percentages of the "paying" creditor's pro-rata share of the debtor's payments to both creditors that are diverted to the other player in her stead. To illustrate, when A employs strategy a1 and B uses strategy b2, A will receive all of her pro-rata share of the debtor's payments plus 10% of B's share. As the game is zero-sum, B pays for A's larger recovery

---

35. *See* RAPOPORT & CHAMMAH, *supra* note 2, at 13–14.
36. *Cf., id.* at 9–13.

by receiving 10% less than her pro-rata share of the debtor's payments to both creditors.

The Two-Creditors' Zero Sum Game
(numbers in percentages of "paying" creditor's pro-rata share)

	B's Strategies			Row Minima
	b1	b2	b3	
a1	−10	+10	+60	−10
a2	−40	−10	+50	−40
a3	−60	−40	−10	−60
Column Maxima	−10	+10	+60	

(A's Strategies label on left side)

The payoffs in the matrix reflect the principle that relative diligence reaps its reward. Each party's first strategy, a1 and b1, therefore consists of more coercive collection measures than his second, and his second more than his third. The least coercive strategy, a3 or b3, may consist of nothing more than the creditor's refusal to grant extensions. In employing his intermediate strategy, a2 or b2, the creditor may supplement this stance with repeated demands for payment. The creditor may act in the most coercive manner, a1 or b1, through threatened or actual seizure of the debtor's property.

A will achieve his best outcome by resort to strategy a1. By that choice, he may recover all of his pro-rata share of the debtor's payments to both creditors plus 60% of B's share. But A's use of his most coercive strategy will only produce this result if B opts for strategy b3. For example, B will capture all of his pro-rata share of payments plus 10% of A's share by employing strategy b1 against A's use of a1.

Upon reflection, any competent player in a game with an experienced opponent will realize that his concern is with the least payoff that he will receive from his choice of any particular strategy.[37] Therefore he will select the strategy that contains the best of these least payoffs. To illustrate this point, matrices for zero-sum games usually provide payoffs in which a

---

37. *See* RAPOPORT & CHAMAH, *supra* note 2, at 21–22; Oskar Morgenstern, *Game Theory: Theoretical Aspects*, in 6 INTERNATIONAL ENCYCLOPEDIA OF THE SOCIAL SCIENCES 62, 64 (1968).

player's choice of strategy to minimize loss differs from the strategy that, in the absence of rational play by his opponent, he would choose to obtain his maximum payoff. Stating the payoffs in The Two-Creditors' Zero-Sum Game in that manner, however, would ignore the usual applicability of the principle of relative diligence: one player scores on the other by using more coercive measures than his opponent.[38] The least payoffs to A for each of his three strategies are contained in the horizontal row minima. Seeking the maximum of the row minima, the "maximin,"[39] A selects strategy a1. On the same basis, B, whose gains are measured by numbers less than zero, is attracted to the minimum of the column maxima, the "minimax,"[40] and he chooses strategy b1.

The point at which the maximin equals the minimax is known as the "saddle point." One theoretician concludes that most games will have no saddle point in pure strategies, such as that posited for The Two-Creditors', Zero-Sum Game. He illustrates such games by "matching pennies," a contest in which A wins only if the side of his coin presented matches that of B's. In this game, A will certainly lose if B is able to predict A's choice of heads or tails. A must employ a chance mechanism to prevent B from doing so, one which selects either side of his coin with equal probability. Such randomization may be achieved by flipping the coin before each showing of it.[41] The importance of relative diligence in most collection contests dictates, however, that a depiction of the typical creditors' contest contain a saddle point in pure strategies, each players choice of his most coercive strategy. Each player will choose the strategy producing this result because he can do no better against a knowledgeable, self-interested opponent. While A will lose 10% of his pro-rata share of the total payments to both creditors to B for A's use of strategy a1 against B's use of b1, A's choice of either other strategy would result in even greater losses against B's use of b1.

The payoff matrix need not reflect one creditor's priority over another in a contest of both's ultimate coercive strategies as a necessary element in a creditor's zero-sum game. But neither is the outcome in the matrix in this respect an aberrant one. It merely reflects a common fact in collection cases: that some creditors, whether based on their secured status or their employment of more imaginative, perhaps one might say even creative, collectors, are able to project a more commanding posture than others.

---

38. *See* RAPOPORT & CHAMMAH, *supra* note 2, at 21–22; Morgenstern, *supra* note 37, at 64.
39. *See* Morgenstern, *supra* note 37.
40. *See id.*
41. *Id.* at 64–65.

Of what utility is the preceding model of the creditors' zero-sum game in informing our understanding of strategic choices in actual collection contests? With what frequency do such contests fit within the theoretical model? To pattern real contests on the model, certain conditions must exist.

Doubtlessly, there are cases where the debtor's distress is beyond relief, where financial failure is imminent, and the sole issue is how distribution from a finite pool of assets devoted to repayment of liabilities will be made among competing creditors. And given the priority generally accorded superior methods of coercion, it seems highly unlikely that any creditor in such a contest would perceive her maximin (or minimax) as dictating anything other than the employment of her most coercive strategy. Thus, there are collection contests that are zero-sum games with saddle points that lie at the intersection of the contestants' most coercive strategies.

The zero-sum model assumes wider applicability to creditors' contests when the limits of the requirement that creditors perceive the contest as one for a limited pool of assets are explored. This characteristic is fundamental in distinguishing zero-sum games from those of Prisoner's Dilemma, for, as the preceding analysis has shown, the basic difference in the types of conflict experienced by the contestants in these two games stems from the futility of their cooperation in the former contest and the promise of that action in the latter one.

Zero-sum games have a satisfactory solution because the interests of the parties are diametrically opposed, whether a saddle point results from the use of pure strategies or random ones. No gain for one player can occur without a corresponding loss for the other, because no concerted action by the players can increase the pool of assets out of which their claims will be paid.

But previous examination of the matrix for Creditor's Dilemma established that the noncooperative equilibrium of mutual coercion provides only the third-best recovery for each party in that contest. In all other outcomes, it is not possible to make one creditor better off without making the other worse off. Significantly, the equilibrium point for a noncooperative game of Creditor's Dilemma is the only one of the four payoffs that is not Pareto-optimal.[42] "[A]n equilibrium is said to be 'Pareto-optimal' if (and only if) there is no possible movement from it that could make *everyone* better off."[43] Both creditors could double their payoffs if they each substituted their cooperative strategies for their coercive ones.

In accordance with this basic distinction in the two types of games, the matrix for The Two-Creditors' Zero-Sum Game assumes that the creditors

---

42. *See* SHUBIK, *supra* note 2, at 217–39.
43. PAUL A. SAMUELSON, ECONOMICS 462 note 12 (10th ed. 1976) (emphasis in original).

cannot increase debtor's aggregate payment on their claims by joint cooperative action in their recovery efforts. This assumption is necessary to pattern the creditors' contest on the model of a zero-sum game. The matrix, however, need not necessarily assume, as it does for the sake of simplicity in presentation, that the debtor's aggregate payments cannot be increased by the unconcerted recovery actions of his creditors.

There are doubtlessly collection cases, probably numerous ones, in which creditors believe that their individual actions will result in some further application of the debtor's resources to payment of their claims. These cases too may qualify as zero-sum games. To do so, however, they must be viewed by creditors as contests with other creditors. If creditors believe that their individual collection efforts will produce payments in an amount sufficient to pay all the debtor's obligations, they will not perceive the collection setting as presenting a contest, and game theory will have no relevancy in predicting their actions. To qualify as zero-sum games, these cases must be ones in which creditors believe that the pool of assets that will be used to pay them is expandable by their efforts, but not to an extent sufficient to pay all the debtor's claims. This chapter will subsequently examine the propensity of creditors to employ ever increasing pressure on debtors in collection cases that are not perceived as contests with other creditors.

What effect on the predictive power of the zero-sum model results from this recognition that creditors may enhance their aggregate recovery by their unconcerted collection efforts? To the extent that the effect is significant, it appears to be a positive one. The unconcerted action that creditors perceive as conducive to increasing the resources devoted to payment of their claims is likely to be more and not less coercive tactics. Thus, the action dictated by this theoretical impediment to modeling collection contests as zero-sum games will normally serve merely to reinforce creditors' choices of their most coercive strategies as predicted by application of a pure zero-sum model that ignores creditors' ability to increase aggregate recovery by unconcerted action.

One further aspect of the requirement that creditors by joint action cannot increase aggregate recovery in a zero-sum game is worthy of note. Although the gross payoffs in The Two-Creditors' Zero-Sum Game afford no possibility of the creditors increasing recovery by agreement among themselves, those creditors may be motivated to cooperate in their efforts when they consider the effects of collection costs on their net recovery. To illustrate this point, the matrix provided for the contest has been designed to reflect the same distribution of the debtor's aggregate gross payments among his creditors from each creditor's choice of his least coercive strategy (a3 and b3) as from each creditor's choice of his most coercive strategy (a1 and b1), the saddle-point equilibrium. If a more coercive strategy is more costly—and usually it will be—each creditor's net recovery could be increased by an agreement among those parties to use their least coercive strategies.

Although this observation qualifies the theoretical application of the zero-sum model in some creditors' contests, it probably has little significance in reducing the model's predictive value in actual cases where creditors will not perceive their payoffs in the precise manner contained in the matrix. But even if creditors have this perception, other considerations appear to diminish the erosive effects of this cost-reduction factor on the applicability of the model.

Creditors will generally have far less incentive to cooperate in choosing collection strategies when they direct their effort solely to the reduction of collection costs than when that effort is motivated by the probability of increasing aggregate recovery from the debtor, as in a Creditor's Dilemma. And as the transaction costs of securing cooperative creditor action often defeat that action in a Creditor's Dilemma, it seems that these costs will pose an even more significant barrier when measured against the smaller benefits at stake in a zero-sum contest. Additionally, creditors' propensity to perceive the pool of assets directed to payment of claims as being expandable by their employment of more coercive strategies will militate against agreements by creditors to employ less coercive strategies merely to curtail collection costs.

To what extent does the zero-sum model reflect the vast amount of collection activity in this country? While no empirical data exists to support my conclusion, I believe this model explains the actions of collectors in far more cases than one based on Prisoner's Dilemma. Why?

Cooperating creditors collect more in a collection case that is properly modeled on a Prisoner's Dilemma because the debtor commits, to a far greater extent than in a zero-sum game, his subsequent earnings and possibly other subsequent income and acquisitions of property to payment of creditors. But the case is properly modeled on a Prisoner's Dilemma only if the debtor is committed to a workout and the workout is feasible. Cases of significant overextensions of credit pose a substantial risk that a debtor will resort to bankruptcy or find other means to immunize assets acquired after creditors get serious in their collection efforts.

Once a debtor's basic-comfort level of current consumption is threatened, each additional dollar diverted from current consumption to debt service will result in greater hardship to the debtor than that resulting from the dollar diverted immediately before it.[44] The increasing marginal costs of these losses in disposable income will tend to defeat the workout efforts of all

---

44. The principle is the converse of the rule of diminishing marginal utility. *See* SAMUELSON, *supra* note 43, at 433–35 (the concept of diminishing marginal utility recognizes that, while increased consumption of a good increases total utility, the extra utility added by each additional unit tends to decrease).

but the most serious-minded and self-disciplined of debtors who are significantly overextended.

The rigors of this elementary economic principle seem most pronounced when a debtor's overextension results solely from volitional expenditures. After having grown accustomed to spending more than his income, he must adjust his present consumption to an amount that is less than that income, often considerably so, in order to fund the workout. But even when overextensions result from non-volitional expenses or temporary suspensions of income rather than a propensity to overconsume, living within the strictures of a workout may exceed the abilities of many debtors.

The zero-sum model does more, however, than explain creditors' use of coercive measures in those many contests that would fall within its boundaries if information were costless and creditors based their strategic choices on an objective assessment of all relevant facts. The very frequency with which it may be properly applied engenders, I suspect, a presumption by creditors that most cases of serious overextension are controlled by the behavior the model predicts. The mind-set of creditors is no less a product of their environment than that of any other occupational group. They are witnesses to far too many financial tragedies to be other than skeptically disposed toward the ostensible benefits of granting extensions to debtors who are already in serious arrears.

To be sure, the presumption in favor of the zero-sum model is rebuttable, but the burden of doing so is placed upon the proponent of creditors' indulgences, upon the proponent of a Creditor's Dilemma model. Some of the difficulties of meeting this burden of proof were previously explored in Kelly's case.

Information supplied solely by the debtor concerning his willingness and ability to successfully complete a workout is suspect, at least for its objectivity if not also for its veracity. While a creditor can often easily verify certain facts relative to the debtor's ability, such as his salary, that creditor may face more difficulty or incur greater costs confirming other data, such as the extent of the debtor's obligations to other creditors. But it is probably not as much the debtor's representations of fact as it is his promises—his willingness to persevere in the workout—that most concerns creditors. What is needed is some basis for assessing the debtor's perseverance, and that often requires a greater investment of time and effort than a single creditor can economically justify.

Even if these obstacles can be surmounted, a final significant one remains. The creditor who contemplates sponsoring the workout must believe that the other creditors can be convinced, as he is convinced, of the relevancy of the Creditor's-Dilemma model. If the other creditors do not share that belief, they will be playing a zero-sum game in contests in which the debtor is seriously overextended and will invariably counter the sponsoring creditor's cooperative strategy with their most coercive ones. But if the sponsoring creditor's own decision in favor of the merits of a workout is a close call—and in many cases it will be—predicting the actions of the other creditors presents the greatest hazard of all.

To the extent that this presumption in favor of the zero-sum model controls, it contributes significantly to understanding creditors' failure to exert greater effort on behalf of workouts in cases where their recovery might be enhanced by the workout. If creditors view the contest not as one of Creditor's Dilemma but as a zero-sum game, they have no incentive to seek mutual cooperation.

The earlier analysis of factors that defeated Kelly's attempt at a workout, one that did not directly invoke the insights of game theory, reached the same conclusion as the foregoing exploration of collection practices that result from misapplication of the zero-sum model. Competition among creditors presents the central problem in overcoming overly coercive methods of collection. Moreover, the need to successfully compete accounts for the dilemma faced by creditors without a cooperative solution to contests that they do recognize as ones of Creditor's Dilemma.

But applying game theory, both zero-sum and Prisoner's Dilemma, to collection cases does more than reinforce the conclusion reached previously in the analysis of Kelly's case. Recognizing creditors' contests as flawed or pure games of Prisoner's Dilemma provides the foundation for exploring solutions to creditors' problems in the larger context of applied game theory's concern with securing beneficial cooperative action in all instances in which the need for cooperative action may arise. Before examining, however, these solutions to the generalized problem of collective choice, another type of creditor misperception of the collection setting deserves recognition for the important role it plays in thwarting beneficial collective action by creditors.

The universe of a creditor's perception of any given collection case is not limited to a choice of the Zero-Sum or Prisoner's-Dilemma models. If there are instances in which a creditor uses coercion because she believes the debtor will soon fail, and she wishes to reap a larger share than her fellow creditors of the limited assets that will be devoted to payment of the debtor's obligations, there are also other instances, probably more numerous ones in the early stages of the collection process, in which a creditor does not perceive her employment of coercive measures as visiting any harm on her fellow creditors. If a debtor has just commenced his slide into profligacy, firm demands, even legal action, by a collector may provide the impetus to turn the debtor around, not only to the benefit of the disciplining creditor but to that of the debtor's other creditors as well.[45]

---

45. *See* 4 ADAM SMITH, AN INQUIRY INTO THE NATURE AND CAUSES OF THE WEALTH OF NATIONS 423 (Edwin Cannan ed. 1937) (1776).

Certainly, initial collection effort must often be guided by the presumption of a need to instill financial discipline in the debtor. But here too, the manner of doing so is not to indulge the debtor's requests for extensions but to demand that defaults be promptly cured.[46] Thus, even before the creditor knows that she is engaged in a contest with other creditors, a pattern of coercive collection has often been established, one which may bias the creditor, once a creditors' contest is apparent, toward applying even more stringent coercive strategies based on the model of a zero-sum game.

Any compendium of the possible misperceptions of a Creditor's Dilemma as an activity not involving a contest with other creditors would be incomplete without one further recognition. In some instances, but probably not many, the creditor's mistaken assessment may lead to her erroneous employment of a cooperative strategy. Creditors often make extensions of consumer debts without regard to whether other creditors are adjusting the terms of their loans. This action will prove advantageous to the creditor granting the extension, as well as the debtor, if the latter's financial condition is not such as to preclude the debtor's financial recovery without concessions from his other creditors. But in a case in which the cooperation of more than one of the debtor's principal creditors is required to avert financial failure, a creditor who mistakenly grants an extension when others do not is vulnerable to their employment of coercive strategies. While asymmetries of information and mistaken judgments may result in some creditors making this error, it seems probable in most cases that by the time the debtor is in serious default, all creditors will perceive the matter as a collection contest and employ coercive strategies as predicted by the preceding analyses of Creditor's Dilemma and zero-sum games.

---

46. *See* COLE, *supra* note 26.

5

Resolving the Creditor's Dilemma

A. Solutions Suggested by Applied Game Theory

(1) Introduction

In the preceding chapter, the search for rational justifications for creditors' failure to cooperate in beneficial workouts led to the recognition that the problem facing creditors in these settings fits within a broader framework of inquiry: the Prisoner's Dilemma concept of game theory. Placing creditors on the Prisoner's Dilemma playing field has served thus far to reinforce conclusions regarding creditor behavior reached independently by earlier exploration of the race-of-diligence motive for coercive collection.

But use of the Prisoner's Dilemma framework serves another and equally important function, which will now be examined. Analysts who recognize creditors as participants in Prisoner's Dilemmas have a broader base for the examination of remedial measures than those analysts who fail to generalize the creditors' problem.

Once the problem of securing beneficial collective action by creditors is recognized as a subset of a broader issue—the problem of obtaining collective action in all the social, economic and political contexts in which joint participation by individuals is required to achieve a common goal—the basis exists for exploring solutions to the narrower issue by application of the revelations of theorists studying the broader one. This is not to suggest that the analysis of collective action in generalized ways—a fairly recent academic industry, whether viewed from the standpoint of the evolution of game theory or that of any other generalized collective decision-making construct[1]—has yielded a comprehensive, verifiable theory for resolving

---

1. Russell Hardin, Collective Action 7, 16–28 (1982).

the problem of collective action.² Certainly that is not so in the myriad substantive areas in which the problem has arisen.³

Russell Hardin observes that "[a]lthough not generalized until recently, the problem of collective action has been recognized in a remarkable variety of contexts over many centuries."⁴ He provides illustrations of the universality of recognition of the problem from such diverse sources as Plato's *The Republic*, Joseph Heller's *Catch-22*, Vifredo Pareto's *The Mind and Society*, and even from the common policy in restaurants of imposing a stated service charge on customers who are part of large groups, whose members are otherwise prone to rely on other members to provide the tip.

Hardin finds that the generalization of the problem of collective action stems from theoretical advances contributed by two major analytical traditions, the theory of public goods and game theory. He attributes the recognition of the theory of public goods to the work of Paul Samuelson.⁵ While Hardin recognizes that public goods are defined by both "jointness of supply," (one person's consumption of the good does not reduce the amount available to others), and "impossibility of exclusion," (relevant people may not be excluded from consuming the good), he notes the contributions of Mancur Olson to expanding the scope of Samuelson's work:

> Olson's analysis of collective action depends not on jointness, but only on the impossibility of exclusion, or more accurately, on the de facto infeasibility of exclusion. For example, if the law says that wage rates in a factory must be uniform for each job category, nonunion employees cannot easily be excluded from enjoying the benefits of union-negotiated wage increases. The central relationship between the analysis of public goods and the problem of collective action, then, is that the costliness or de facto infeasibility of exclusion from consumption of a collectively provided good usually eliminates any direct incentive for individual consumers to pay for the good.⁶

---

2. *See id.* at 14 ("To attempt a complete decision theory of human behavior would be absurd—certainly at this time, but probably also in principle.").

3. Both Robert Axelrod's and Russell Hardin's seminal studies are replete with numerous examples of the problem of collective action. Their illustrations cross otherwise unrelated substantive fields and disciplines, range in number of participants from actions by small groups to mass movements, and encompass problems that vary in magnitude from the trivial to the most significant. *See* ROBERT AXELROD, THE EVOLUTION OF COOPERATION (1984); HARDIN *supra* note 1.

4. *See* HARDIN, *supra* note 1, at 7–9.

5. *See id.* at 16 & n.1, (*citing* Paul A. Samuelson, *The Pure Theory of Public Expenditure*, 36 REV. ECON. & STATISTICS 387 (1954)).

6. *Id.* at 19–20 (*citing* MANCUR OLSON, JR., THE LOGIC OF COLLECTIVE ACTION (1965)).

After Hardin recognizes the extension of the theory of public goods to Olson's theory of collective action, he demonstrates that Olson's theory and the theory of Prisoner's Dilemma are equivalent by constructing an "individual vs. collective" game matrix.[7] Hardin concludes that "the vast body of experimental and theoretical work on Prisoner's Dilemma is relevant to the study of collective action in general and conversely that the growing body of work on collective action can be applied to the study of the Prisoner's Dilemma."[8]

But prior failure to generalize the problem of collective action accounts for the various guises in which it appears in various disciplines and contexts. Hardin himself is unable to resist the temptation to clothe the problem in new dress and seizes upon Adam Smith's principal of the invisible hand to do so. He notes that while collective interest is often best served by private interest-seeking in market exchange, self-interest can prevent parties from succeeding in collective endeavors. Thus, "all too often we are less helped by the benevolent invisible hand than we are injured by the malevolent back of that hand."[9] In addition to "the back of the invisible hand," Hardin recognizes "the free rider problem" and "the condition of common fate" as other appellations for the problem of collective action.[10] Recall from Chapter Three, Douglas Baird and Thomas Jackson's "common pool" analogy, which vividly depicts the very problem with which this study is concerned. They liken the problem of too many creditors chasing too few assets to that of too many people fishing from a common pool: in both instances there is a disproportionate sacrifice of future for present value.

The insights of these theorists contribute to an understanding of why a systematic resolution of the Creditor's Dilemma short of bankruptcy has required fashioning a unique form of private-sector agency, Consumer Credit Counseling, to address the issue. I look first at possible solutions to Creditor's Dilemma not employing this agency. The final part of this chapter addresses the unique role credit counselors play in providing the only systematic, private-sector resolution extant of the Creditor's Dilemma.

(2) Tacit Reciprocity

The exploration of generalized solutions to the problem of collective action to discover a resolution of the Creditor's Dilemma need not com-

---

7. *Id.* at 25–28.
8. *Id.* at 28–29.
9. *Id.* at 6.
10. *See id.* at 7.

mence, however, with the abstractions of game theorists that are fashioned from their study of complex problems in disciplines such as political science, economics, psychology and sociology. The work of Robert Axelrod[11] suggests the possibility of a simpler resolution of the Creditor's Dilemma. He has found that if a certain condition exists, cooperation may develop in even pure games of Prisoner's Dilemma where the parties are offered no means of resolving their conflict by explicit agreement. And this condition, which offers humankind such promise, requires nothing more than that the participants engage each other not in single plays of Prisoner's Dilemma, where empirical data supports the theoretical equilibrium of mutual failure to cooperate,[12] but in repeated plays of the game.

Before Axelrod's work, game theorists had recognized that these iterated games, which necessarily incorporate the element of strategic interaction over time, were of fundamental importance in motivating cooperation by the opportunity they presented for tacit communication.[13] That recognition did not immediately yield, however, a theoretical answer to the question of whether iterated games would actually produce mutual cooperation and, if so, under what conditions.

To explore these questions, consider the concept of a complete strategy. A complete strategy states what the player will choose for each separate move in the game or provides a rule from which the player's choice for the move can be determined.[14] The question of whether iterated games will result in mutual cooperation may now be put in another form: what choice of complete strategies works best for a player against all choices of complete strategies by the other player in the game?

Before addressing that question directly, a threshold issue that may determine whether some iterated games provide any motivation for cooperation should be examined. If an iterated Prisoner's Dilemma is played a finite number of times and that number is known by the players, some game theorists posit that they will have no more incentive to cooperate than in a single-play Prisoner's Dilemma. On the last play, each will employ her noncooperative strategy because a cooperative one on that play will not influence future moves of the other player. But that being so, on the next to last move neither will cooperate because each knows what the other will do on the last one. Hence, the game unravels all the way back to the

---

11. AXELROD, *supra* note 3.
12. *See* Gerrit Wolf & Martin Shubik, *Concepts, Theories and Techniques: Solution Concepts and Psychological Motivation in Prisoner's Dilemma Games*, 5 DECISION SCI. 153 (1974).
13. *See* HARDIN, *supra* note 1 at 145.
14. *See id.* at 149.

first play in which neither player cooperates.[15] This unraveling theory, however, is not universally accepted:

> That this paradox is not to be taken seriously can be intuited quickly enough if we imagine being placed before *exactly 1 million* or, alternatively, before *about 1 million* plays of Prisoner's Dilemma. Would any sophisticated human being really behave so differently in these two situations as always to defect in the first and always (or almost always) to cooperate in the second?[16]

Whether the unraveling theory has validity in some instances is an issue that is not relevant, however, in examining many of the iterated games that occur in real-life settings in which the players, nor anyone else for that matter, know the number of plays in the game. The same firms of creditors frequently encounter each other in collection cases, but unless one of them anticipates retiring from the field of debt recovery after engagement in a certain number of additional collection contests, none of the firms will be engaged in a play of Creditor's Dilemma with a finite number of moves. Firms of creditors may change owners or managers, but they seldom abandon a well established business. Thus, the unraveling theory need not concern us in our effort to determine a player's choice of a complete strategy that works best against all choices of complete strategies by the other player in the game.

To illustrate that there is no dominant complete strategy in an iterated game of Prisoner's Dilemma[17], and therefore that neither party will necessarily be motivated to cooperate on any separate play of that game, consider a simple iterated Prisoner's Dilemma consisting of 10 moves. To avoid any complications arising from the unraveling theory, however, assume that the game is of unknown duration to the players. Each of the two players must choose one of only two complete strategies offered them.

One of these complete strategies requires the player using it to always defect. Thus, that player will use the noncooperative strategy on every move in the game regardless of what the other player has done on previous moves. Game theorists customarily use the term "defection" to describe the noncooperative strategy in a Prisoner's Dilemma. In presenting the matrix for Creditor's Dilemma, I substituted the term "coercion" for "defection" as explained in Chapter Four. In discussing Prisoner's Dilemma in this section of the text, however, I use the conventional terminology.

---

15. *See* R. Duncan Luce & Howard Raiffa, Games and Decisions 94–102 (1957).
16. Hardin, *supra* note 1, at 146 (emphasis original).
17. *See id.* at 149–50.

The other complete strategy, known as "TIT FOR TAT" in the parlance of game theorists, requires that the player employing it cooperate on the first play and on succeeding plays do what her opponent did on the preceding play.[18] In a contest with an opponent using the TIT FOR TAT complete strategy, a player who also chooses TIT FOR TAT will do better than one who chooses to always defect. This results because the sum of 10 payoffs of the reward for mutual cooperation (R) will exceed the sum of one temptation payoff (T) and nine payoffs of the punishment for mutual defection (P).[19] But if an opponent chooses a complete strategy of always defecting, a player will do better by foregoing TIT FOR TAT and also choosing the complete strategy of defection on every play. Here the sum of 10 payoffs of the punishment for mutual defection (P) will exceed the sum of one sucker payoff (S), obtained by cooperating when the opponent defected on the first move, and nine payoffs of the punishment for mutual defection (P). To avoid the problem of discounting the value of future payments, both examples assume that the payoff parameters for later plays of the game are the equivalent of those for the first play.[20]

This simple illustration shows that no complete strategy dominates all others. One such strategy, however, may produce a better result for its user than any other when used in a large number of iterated games with different players who freely formulate their own complete strategies. A better result for the user of this strategy is not measured by whether the complete strategy she employs produces a higher score for her than her opponent in more individual contests than another complete strategy would have done. Instead, the result is measured by the sum of all that player's scores with each opponent she engages in separate iterated games. Whether such a complete strategy exists in a meaningful context and, if so, what it might be, are the questions that Robert Axelrod sought to answer in a highly structured, empirical way.

Axelrod seized upon the novel idea of a computer Prisoner's Dilemma tournament that would pit each entrant's complete strategy against every other entrant's complete strategy. The rules also provided that each entrant's complete strategy would be paired with its own twin and with a complete strategy that cooperates or defects with equal probability.[21] To score the tournament, in which players would engage each other one at a time, Axelrod provided a payoff matrix to be applied to each move in each

---

18. *See id.*
19. *See* Chapter 4 Section (A) for an explanation of the rank order and relative weight of payoffs in a Prisoner's Dilemma.
20. *See* AXELROD, *supra* note 3, at 12–13.
21. *See* id. at 29–31.

game.[22] The goal of each player was to attain the highest sum of these payoffs for individual moves in all games. No additional points were awarded for attaining a higher aggregate score for all moves in an individual iterated game than the other player in that game. Thus, the winner of the tournament would be the player who submitted the complete strategy that interacted best with all the other complete strategies that were submitted.

To ensure state of the art complete strategies, Axelrod recruited his contestants from game theorists[23] in various disciplines who were familiar with the Prisoner's Dilemma. Five disciplines (economics, mathematics, political science, psychology and sociology)[24] were represented by the contestants who submitted the 14 entries. The use of a computer permitted Axelrod to place no limitation on the complexity and sophistication of the complete strategies submitted.[25] For example, the complete strategy of one entrant estimated the probability that the other player would cooperate after he cooperated and also the probability that the other player would cooperate after he defected. These estimates were updated after each move. This complete strategy then selected the strategy for its next move that would maximize the long-term payoff of its user against that player.[26]

When each game of 200 moves had been run,[27] Axelrod found, to his "considerable surprise," that "the winner was the simplest of all the programs submitted, TIT FOR TAT. Recall that TIT FOR TAT is merely the strategy of starting with cooperation, and thereafter doing what the other player did on the previous move."[28]

Recognizing that a single tournament was not definitive, as the effectiveness of a particular complete strategy depends on the nature of the other complete strategies with which it must interact,[29] Axelrod conducted a second round of his tournament. He described the second round as "a

---

22. *See id.* at 30–31 (Axelrod's matrix assigned 3 points for mutual cooperation and 1 point for mutual defection. Where the players' choice of strategies did not coincide, the defecting player received 5 points and the cooperating one received 0 points.).

23. *See id.* at 30 ("If the participants are recruited primarily from those who are familiar with the Prisoner's Dilemma, the entrants can be assured that their decision rule [complete strategy] will be facing rules of other informed entrants.").

24. *Id.* at 31.

25. *See id.* at 30 ("The program [complete strategy] has available to it the history of the game so far, and may use this history in making a choice.").

26. *See id.* at 34–35.

27. *See id.* at 30. "[T]he entire round robin tournament was run five times to get a more stable estimate of the scores for each pair of players. In all, there were 120,000 moves, making for 240,000 separate choices." *Id.* at 31.

28. *Id.* at viii.

29. *Id.* at 40.

dramatic improvement over the first round" in that the second-round entrants "were given the detailed analysis of the first round" and "sixty-two entries from six countries"[30] were submitted. The second round was conducted in essentially the same manner as the first one,[31] and after the results of "over a million moves" had been tallied, TIT FOR TAT was again the winner.[32] Once more, that complete strategy was the simplest submission for the round. The basic principle of TIT FOR TAT is simply that of reciprocity.[33]

That the results in both rounds of the tournament had significant import was readily apparent to Axelrod. "Something very interesting was hap-

---

30. *Id.* at 41. "The contestants ranged from a ten-year old computer hobbyist to professors of computer science, physics, economics, psychology, mathematics, sociology, political science, and evolutionary biology. The countries represented were the United States, Canada, Great Britain, Norway, Switzerland, and New Zealand." *Id.*

31. *See id.* at 42. Minor end-game effects in the first round were eliminated in the second by announcing in the rules that "the length of the games was determined probabilistically with a 0.00346 chance of ending with each game move." *Id.* at 42. For theorists' views of the effect of players knowing when the last move will occur, *see supra* notes 15–16 and accompanying text.

32. *See* AXELROD, *supra* note 3, at 42.

33. While Axelrod recognizes the principle of reciprocity as the foundation of TIT FOR TAT, his analysis of that complete strategy is a more detailed one that addresses why it prevailed over others that also embraced the principle of reciprocity. Axelrod attributes the success of TIT FOR TAT to "its combination of being nice, retaliatory, forgiving, and clear." *Id.* at 54.

Of these four properties, he found that of being nice, meaning never being the first to defect, was the one that best distinguished the relatively high scoring entries in the tournament from the relatively low scoring ones. *Id.* at 33–36, 43–44. In addition, TIT FOR TAT is "not very exploitable," *id.* at 32, because it is also retaliatory, that is "it immediately defects after an 'uncalled for' defection from the other [player]." *Id.* at 44. Axelrod concludes that "unless a strategy is incited to an immediate response by a challenge from the other player, the other player may simply take more and more frequent advantage of such an easygoing strategy." *Id.* But TIT FOR TAT combines this short-term retaliation with a long-term policy of forgiveness. "[I]t forgives an isolated defection after a single [defection] response." *Id.* at 46. While the proper balance of retaliation and forgiveness depends upon the other complete strategies present in a given environment, Axelrod notes that "the evidence of the tournament suggests that something approaching a one-for-one response to defection is likely to be quite effective in a wide range of settings." *Id.* at 120.

Finally, because the first three properties of TIT FOR TAT are easy to recognize, the fourth property of that strategy exists. "[I]ts clarity makes it intelligible to the other player, thereby eliciting long-term cooperation." *Id.* at 54. Too complex strategies cannot be distinguished by others with which they interact from a purely random strategy, and this may result in some of the interacting strategies (but not TIT FOR TAT) defecting on all moves after the misassessment is made. *See id.* at 122.

pening here. I suspected that the properties that made TIT FOR TAT so successful in the tournaments would work in a world where *any* strategy was possible. If so, then cooperation based solely on reciprocity seemed possible."[34] From this genesis, Axelrod's significant contribution to game theory consists of two proofs: (1) how reciprocity based cooperation can emerge in a predominately noncooperative environment and (2) how, once established, it can resist invasion by those using a less cooperative strategy.[35]

As Axelrod proved that in some instances reciprocity alone may trigger cooperation, his findings are germane to any comprehensive quest for the resolution of the Creditor's Dilemma. Certainly creditors that operate in the same geographic area and have a substantial number of claims meet each other repeatedly in collection cases, and, thus, they engage one another in iterated Creditor's Dilemmas. Furthermore, the large number of encounters in a continuous stream among these creditors seems to satisfy a major requirement that Axelrod imposes on his reciprocity-based model of cooperation. For TIT FOR TAT to pervade a community, its members must not unduly discount the value of future payoffs relative to present ones.[36] This discounting may result from too great a "likelihood that the interaction will end soon," or too much "preference for immediate benefits over delayed gratification, or to any combination of these two factors."[37] Neither factor causing discount of future payoffs should normally be present in games of Creditor's Dilemma.

Where Axelrod's model of reciprocity applies to a workout that all creditors perceive as a Creditor's Dilemma, transaction costs that arise solely from the need to obtain the cooperation of creditors should pose no barrier because creditors should cooperate based simply on conventions that arise from their previous interactions.

Nevertheless, serious problems arise in applying Axelrod's model to typical consumer-credit collection cases. As previously observed, creditors must frequently be convinced of the need for extensions and the benefits of granting them. Unlike the players in Axelrod's tournaments, someone must convince these creditors that they are engaged in a game of Prisoner's Dilemma. Obviously, Axelrod's model of cooperation based simply on the convention of reciprocity does not address the need to find someone willing to incur the transaction costs presented by this threshold problem.

---

34. *Id.* at viii.
35. *See id.* at 55–69, 206–15.
36. *See id.*
37. *Id.* at 128.

Axelrod's model also appears to be inapposite in the rare case when the benefits to aggregate recovery of cooperation in a workout appear obvious to all creditors and each creditor is aware that the benefits are obvious to all. Axelrod's players knew who they were engaging and how the other player responded in each move of each game. Reciprocity in his model rested on each player's response to another's prior choices. Axelrod concluded that "[w]ithout this ability to use the past, defections could not be punished, and the incentive to cooperate would disappear."[38] Unfortunately, creditors in typical consumer-credit collection cases do not interact directly but act only through an intermediary, the debtor. While in some instances the debtor will inform his creditors of other creditors' actions, this information is frequently lacking or suspect. For instance, the debtor may attempt to obtain one creditor's extension by misrepresenting that other creditors have agreed to extend their payment terms.

Another factor probably has even greater significance in dismissing the relevancy of Axelrod's tacit-reciprocity model as the basis for a systematic resolution of the Creditor's Dilemma. Each of Axelrod's players participated in an iterated game with every other player, but each did so one player at a time.[39] To replicate this aspect of the tournament, no more than two creditors, who had claims of sufficient size to affect the debtor's economic viability, could be engaged in any move to collect from that debtor. Although Kelly's case presented just such a two-person Prisoner's Dilemma, her case is probably atypical in this age of numerous issuers of credit cards and other extenders of credit. Because any creditor may use a defection strategy to score on other creditors who cooperate in a workout, a creditor contemplating cooperation must necessarily concern himself with the strategic choices of all other significant creditors in the game. Even if a creditor somehow knows the identity and past practices of these other creditors, an increase in the number of players can only increase his chances of encountering one whose past moves dictate his choice of a noncooperative strategy for his present move.

---

38. *Id.* at 182. Although each player in the tournament had perfect knowledge of every other player's prior choices, Axelrod recognizes that "[i]n many settings…a player may occasionally misperceive the choice made by the other." *Id.* He reports that he ran a modified first round of the tournament again to explore the implications of such misperceptions. *See id.* at 182–83. In the modified round, "every choice had a 1 percent chance of being misperceived by the other player." *Id.* at 183. Even this small level of misperception "resulted in a good deal more defection between the players." *Id.*

39. *See id.* at 30–31.

(3) Explicit Methods of Resolution and the Multi-Player Aspect of Creditor's Dilemma

Recognizing that Creditor's Dilemmas are often more than two-party games considerably diminishes the chance of resolving them by cooperation based on tacit reciprocity. But this recognition of the n-party aspect of many Creditor's Dilemmas does much more. Enlarging the circle of assent that is necessary to obtain the requisite amount of cooperation for a workout in a Creditor's Dilemma significantly increases the transaction costs of all methods of resolution, explicit as well as tacit.

The costs to any creditor of getting together with others and convincing them of the benefits of concerted extensions will obviously rise as the number of these creditors increases. But the increment to transaction costs by growth in group size may increase for the following reason as well.

Increasing the aggregate recovery of creditors by their acceptance of a workout plan is a probability but not a certainty in any given Creditor's Dilemma. Because the debtor may fail before completion of the workout, a creditor has reason to prefer that his claim be paid sooner than that of other creditors. The matrix provided earlier for the Creditor's Dilemma reflects the possibility that the debtor will fail before completion of the plan by providing no payoffs that reflect full payment of either creditor's claim. A creditor can seek early payment, without increasing the debtor's aggregate periodic payments and thus without undermining the debtor's economic viability, only if other creditors will agree to accord him a greater share of the debtor's initial payments than they do themselves until he is paid in full.

The presence of a large group of creditors increases the bargaining power of a single holdout to insist on receiving this expedited payment of his claim as the price of his promise to forego a choice of his coercive strategy. When the number of other creditors is large enough, the price of a holdout's preference to any one of them is less likely to diminish the value of the cooperative solution to a point where it is less attractive than recovery under the race of diligence. The holdout then has reason to believe that he may successfully divert a greater amount than his pro-rata share of early payments to his claim.

Obviously, however, other creditors will have like incentive to cast themselves in the role of the holdout, a "freerider" on group effort.[40] Thus,

---

40. The problem of "freeriding" is a pervasive one in structuring economic activity. *See, e.g.*, Saul Levmore, *Monitors and Freeriders in Commercial and Corporate Settings*, 92 YALE L.J. 49 (1982). Hardin recognizes "the free rider problem" as another name for the generalized problem of collective action. *See* HARDIN, *supra* note 1, at 7.

creditors will have to engage in more bargaining to placate these holdouts. As a result, cooperation will either be defeated or made more costly.[41]

Mancur Olson has addressed the relationship between group size and the probability that a collective good will be provided by its members. In applying Olson's analysis to creditors in collection contests, consider first the ideal group size, one that ensures that a desired collective good will be provided:

> The smallest type of group—the group in which one or more members get such a large fraction of the total benefit that they find it worthwhile to see that the collective good is provided, even if they have to pay the entire cost—may get along without any group agreement or organization.[42]

Where the benefit of a collective good to a single member of the group is more than her cost, Olson classifies the group as "privileged."[43] At least one of its members has sufficient incentive to see that the collective good is provided.[44] In collection cases, forbearance by a single creditor in the collection of her claim may enable the debtor to successfully complete a workout. A creditor may exercise that forbearance by placing a moratorium on the collection of her claim until other creditors are paid. When a debtor needs more assistance, a cooperating creditor may make a consolidation loan to the debtor, which removes other creditors from the playing field. While instances of one creditor assuming a disproportionately large, or even the sole risk, of a workout exist, the probability of one creditor's benefits from the workout exceeding all the risks of the workout is not likely once the debtor's condition has deteriorated to a point that requires a basic restructuring of the debtor's obligations.

Contrast Olson's privileged group, in which members achieve a collective good without group organization or coordination, with his second type of group, the "intermediate" one. In that group, "no single member gets a share of the benefit sufficient to give him an incentive to provide the good himself."[45] But as the intermediate group "does not have so many members that no one member will notice whether any other member is or

---

41. *See* MANCUR OLSON, JR., THE LOGIC OF COLLECTIVE ACTION 40–41 (1965) (in groups in which there must be 100 percent participation to achieve group-oriented action, the incentive to holdouts makes such action less likely or more costly than it otherwise would be).
42. *Id.* at 46. For the formal proof of Olson's findings, *see id.* at 22–33.
43. *Id.* at 49–50.
44. *Id.*
45. *Id.* at 50.

is not helping to provide the collective good,"[46] the possibility exists of obtaining the good through "some group coordination or organization."[47] Olson's third type of group, the "latent" group, shares one characteristic with the intermediate group: cost-benefit analysis precludes provision of the collective good by any individual member of the group. But in the latent group the possibility of obtaining the collective good through group coordination or organization is also foreclosed "since no one in the group will react if...[another member] makes no contribution."[48]

Having established that creditors do not commonly fall within Olson's privileged group, in which of the other two categories of his typology do they belong? As Olson places only very large groups in his latent category,[49] it would appear that a group of creditors in a typical consumer-collection case must be subsumed within his intermediate category.

This classification is bolstered by Russell Hardin's refinement on the function of group size as a determinant of collective action. He reasons that where the benefit of a collective good is large in relation to its cost, it may pay for some fraction of the group to provide it, even though the members of that subgroup receive no assistance from other members of the group.[50] In the context of a collection case, this means that some of the debtor's unsecured creditors may agree to forego strict proration of the debtor's early payments in favor of others until these preferred unsecured creditors are paid. Members of the subgroup furnishing the collective good by taking delayed payments either view the probability of the workout being successful more favorably than the other unsecured creditors or the preferred creditors are simply better negotiators.

Reasons are usually advanced, however, other than superior negotiating skills, for preferring some creditors over others in the workout. For instance, administrative convenience and cost reduction may justify paying small claimants first, rather than sending them numerous small remittances over the entire term of the workout. Without some justification for preferences, some creditors who would otherwise support the subgroup's plan by extending their terms of payment will be influenced by their perceptions of inequities and withhold their support. This erosion of support for the workout may result in such a decrease in the size of the subgroup

---

46. *Id.*
47. *Id.*
48. *Id.*
49. *Id.* ("the analog to atomistic competition in the nonmarket situation is the very large group").
50. *See* HARDIN, *supra* note 1, at 40–41.

as to preclude its sponsoring of the plan. Where such erosion does not occur, however, agreement by fewer than all creditors in the group may suffice to provide the collective good.

Placing creditors in typical consumer collection cases in Olson's intermediate group means that, while no single creditor will assume all the costs and risks of a workout, there is a possibility that all or some of the creditors may do so. Application of Olson's typology predicts that the collective good of increased aggregate recovery is not beyond attainment provided there is some type of group coordination or organization. Indeed, as a purely theoretical construct, Olson states that in an intermediate group "a collective good may, or equally well may not, be obtained."[51]

Absent a highly specialized type of organization to obtain cooperation among creditors, however, I do not believe that the probability of obtaining the cooperation of creditors in a reasonably promising workout of a consumer debtor remotely approaches even odds. In the types of collective action with which Olson is concerned, group members harbor no doubts about the desirability of the collective good. As previously detailed, however, collection cases pose the threshold issue of whether the debtor will use any extensions given to increase the aggregate recovery of creditors. What is needed in collection cases is not just some effective means of apportioning the costs of obtaining an admittedly desirable collective good. Additionally, some means must be provided to reasonably assure the debtor's creditors in any given case that extensions are in fact desirable. That creditors themselves have fashioned an agency, Consumer Credit Counseling, that effects these ends is testimony to their special needs in resolving the problem of collective action.

How did this agency arise? That question is not answered simply by noting that the number of creditors in most consumer collection cases does not preclude some type of group coordination or organization.

In the subsection that follows, the economies of scale that are effected by using credit counseling in the many cases in which it is employed will be given credit for the success of counseling as an instrument for obtaining collective action in individual consumer collection cases. To attain these scale economies, a significant number of the creditors in all consumer collection cases must accept workouts proposed by credit counselors. Thus, the agency that contains transaction costs for intermediate size groups in individual collection cases could not function without the acceptance and support of its role by a vastly larger group, the general community of creditors.

---

51. OLSON, *supra* note 41 at 50.

Observation of this fact does not alter the size of the relevant group whose cooperation must be obtained in a typical consumer collection case. That group remains where it has been properly placed, in Olson's intermediate-size category. Obviously, the size of the group in any given collection case is limited to those creditors with claims against the debtor in that particular case. But recognition that the facilitating agency needs large-scale creditor support to function efficiently does raise the question of how an agency with that support originated.

Olson has recognized that even in large, latent groups a collective good may be provided as a by-product of some other purpose for which there is sufficient incentive to produce group organization.[52] An organization, like the American Medical Association, offering its members recreational benefits and non-collective products, such as insurance and special journals, may obtain support for a collective good, such as lobbying.[53] Similarly, this by-product theory might explain why creditors cooperate in the collective good of workouts sponsored by counseling agencies affiliated with the National Foundation for Consumer Credit. The Foundation offers its member creditors informational and other services in addition to providing oversight of the work of affiliated counseling agencies.

In the credit-counseling context, however, the by-product theory of collective action does not provide an adequate explanation of creditor behavior. First, the universe of creditors who accept the services of counseling agencies is much larger than the membership of any trade association of creditors. Secondly, an historical argument arises. The first consumer counseling agencies were organized by local creditors specifically to address the problem of workouts. These agencies were not sponsored by any national organization that creditors may have joined for other services that it provided. Today, a counseling agency's affiliation with the National Foundation for Consumer Credit no doubt significantly enhances its credibility with creditors in distant cities and even local creditors not personally familiar with the quality of the agency's work. Moreover, virtually all non-profit consumer counseling agencies in this country now have affiliation with the Foundation. Nevertheless, the credit counseling movement is probably best described as the product of a need perceived and remedy implemented by those actually engaged in collection work.

---

52. *See id.* at 132–41.

53. *Id.* Olson argues "that the main types of large economic lobbies—the labor unions, the farm organizations, and the professional organizations—obtain their support mainly because they perform some function besides lobbying." *Id.* at 135. *See also* HARDIN, *supra* note 1, at 31–35 (examining the merits and shortcomings of the by-product theory of collective action).

B. The Functions of Consumer Credit Counseling Agencies and How Those Functions Differ from Those of Bankruptcy Courts in Chapter 13 Proceedings

Consumer Credit Counseling serves the now familiar creditors' needs of establishing the probable benefit of a concerted extension plan and monitoring both the debtor's and each creditor's adherence to that plan once it is in place. The distinctive feature of credit counseling as a private-sector remedy is that it accomplishes these goals within the limits imposed by the severe cost constraints that distinguish consumer collection cases from large commercial ones.

To be sure, the costs of effecting a workout are generally far greater in commercial than in consumer cases. The transaction-cost barrier, however, is measured by costs relative to benefits. Consequently, the amount of some creditors' claims in commercial cases, normally those of the debtor's principal secured lenders, are often sufficient to justify those creditors assuming the costs of effecting a workout. While the mortgagee of a consumer debtor's home or even the creditor with a security interest in the debtor's automobile may have a relatively large claim, they ordinarily do not have the same incentive to sponsor a workout and forego foreclosing on collateral that their commercial lending counterparts do. Unsecured creditors often lack sufficient financial incentive to incur the costs of working extensively with the debtor in large commercial cases.[54] *A fortiori*, that incentive is largely missing in consumer cases. In short, credit counseling affords economies of scale—in this instance, a reduction in the transaction costs of effecting a promising workout[55]—that other private sector initiatives, such as a single creditor's sponsorship of a plan, do not.

By working with distressed debtors day in and day out, credit counselors hone their skills in assessing whether debtors need extensions and whether they will use them to increase the aggregate recovery of creditors. The repetitive nature of the credit counselor's work provides the special-

---

54. *Cf.* Lynn M. LoPucki, *The Debtor in Full Control—Systems Failure Under Chapter 11 of the Bankruptcy Code?* (pts. 1 & 2) 57 AM. BANKR. L.J. 99, 247 (1983) (an empirical study of unsecured creditors' failure to monitor Chapter 11 reorganization proceedings).

55. *See* C. E. FERGUSON & J. P. GOULD, MICROECONOMIC THEORY 208–09 (4th ed. 1975) (exploring the reasons for economies of scale in the production of goods).

ization and division of labor upon which economies of scale are often based.[56] In this function, the credit counselor plays the role of the prosecutor in the prototype of the Prisoner's Dilemma. She informs creditors of the payoff parameters, obviously not with the precision of the prosecutor but in a manner sufficient to qualify the case as a game of Prisoner's Dilemma. Her representation that the payoffs comply with the two rules of inequality that define the game and that therefore the players' aggregate recovery may be increased by mutual cooperation is implicit in her decision to sponsor the workout.

By also working with the principal creditors in the community on a daily basis, the counseling agency, and often the individual credit counselor, acquire the trust and confidence of these creditors in the agency's role of monitor in Creditor's Dilemmas. Creditors know that the agency will inform on those of them who yield to the lure of the temptation payoff and that it will dismiss the plans of debtors who are using extensions for ulterior motives. In this function, the scale economies are most significant, for once a creditor's trust is established, the cost factor of securing his cooperation is significantly reduced if not removed from all subsequent cases.

While individual credit counseling agencies first earned the trust of creditors, agencies affiliated with the National Foundation for Consumer Credit now benefit significantly from that affiliation. A newly formed agency that has received the approval of the National Foundation may trade on the goodwill of long established agencies in establishing its own reputation. And established agencies dealing with geographically remote creditors may trade on the reputations of agencies that operate in that creditor's usual trade area.

A national association apparently fulfills another need of credit counseling agencies. At their national meeting a few years ago, many credit counselors complained of the collection practices of one large money-center bank that issued credit cards nationally. They informed their national officers that this bank continued to use coercive collection practices following its notification that a debtor had entered into a workout plan under their auspices. Credit counselors subsequently gave this bank excellent marks for cooperating in the workouts they sponsored. Apparently, moral suasion by officers of the national association was needed to bring the recalcitrant bank around.

My impression, however, is that once the cooperation of creditors in the work of counseling agencies is established, those creditors usually remain steadfast in their support of agency-sponsored workouts. In the im-

---

56. *See id.*

agery of Prisoner's Dilemma, removing the wall that separates the prisoners and prevents mutual cooperation among them need only be done once. On subsequent arrests, the prisoners must be placed in the same cell.

The discussion that follows elaborates on the two principal functions of credit-counseling agencies—assessing whether the debtor is a good candidate for a workout and monitoring any workout it sponsors. It also contrasts their role with that of another important means of obtaining collective action in workouts, Chapter 13 proceedings in the bankruptcy courts.[57] These court proceedings terminate the race of diligence by application of the automatic stay of any action to enforce a claim outside the bankruptcy court.[58] They therefore effect cooperation by imposition of the authority of government and are viewed by some as a less troublesome, less costly substitute for out-of-court workouts by consumer debtors. I will argue, however, that, from the perspective of how the two processes are often used in this country, the work of counseling agencies serves a different purpose for a different constituency than debt adjustment in Chapter 13.

While the debt adjustment practices of the various counseling agencies differ in some respects, there appears to be a high degree of uniformity in the work of those affiliated with the National Foundation for Consumer Credit. This uniformity contributes significantly to creditors' acceptance of the work of the affiliated agencies and therefore greatly enhances their efficiency. In addition to acting as intermediaries in workouts, credit counselors assist consumers in two other ways. They assist individuals in budgeting their income and instruct various groups on the proper use of credit. As these agencies are by far the most significant private-sector response to the problem of overextensions by consumers,[59] the outline that follows

---

57. 11 U.S.C.A. §§ 1301–1330 (West 1993 & Supp. 1996).
58. *See* 11 U.S.C.A. § 362 (West 1993 & Supp. 1996).
59. Observing the role of credit counseling a quarter-century ago, Carl Felsenfeld noted the paramount status, even at that time, of agencies affiliated with the National Foundation for Consumer Credit. Following his analysis of agencies that perform debt adjustment for profit, he found:
> Alongside the regulated business and of more recent origin in the field are offices of various sorts that also do debt adjusting and pro-rating and advising of consumers on a non-profit basis. The Office of Economic Opportunity has set up a few pilot offices in various states. Legal Aid offices here and there have adopted this function and perform this specific service. Family Service agencies in various cities throughout the country have also set up debt adjusting on a non-profit basis. However, by far the most significant in recent years has been the growth of the Community Credit Counseling Service offices that exist in fairly substantial numbers around the country. This is an effort that was spearheaded originally by creditor groups who recognized the problems of growing consumer debt and the need for counseling in communities.

addresses their practices as intermediaries in workouts and not those of other private-sector debt adjusters.[60]

Because the source of debt problems for many consumers is a proclivity to defer payments on obligations until ever expanding indebtedness

---

> By and large, the original consumer credit counseling agencies were financed and sponsored by creditor organizations. Stemming from this function, the National Foundation for Consumer Credit, headquartered in Washington, has taken the lead in sponsoring and stimulating the growth of these consumer credit counseling organizations. These are continually growing in number and vitality and have an active and productive national organization.

Carl Felsenfeld, *Consumer Credit Counseling*, 26 BUS. LAW., 925, 929–30 (1971).

60. In addressing solely the practices of agencies affiliated with the National Foundation for Consumer Credit, I give no consideration to the practices of debt adjusters who render their services for a profit. Writing in 1971, Carl Felsenfeld addressed the work of these fee-generating adjusters:

> Credit counseling cannot be thought of except in terms of its relationship to an existing business of arranging consumer debts for payment to creditors. This is a well established business which exists in many forms throughout the country as commercial ventures. They are variously called commercial pro-raters or commercial debt adjusters, and their business is advising consumers as to their financial plight, arranging for settlement of their obligations in some way and, normally, collecting money from them and paying it to their creditors as a method of working out the debts—all for a fee.

Id. at 926–27.

But Felsenfeld found that the business of debt adjustment had been subject to a great deal of criticism, essentially because those who practiced it made money as a result of other's financial woes. *See id.* at 927. He noted other causes as well, including instances of false advertising, excessive fees, creditors' reluctance to deal with commercial adjusters and adjusters that absconded with the plan's funds. *See id.* at 927–28. Commenting on the state of legislative reactions to these criticisms some 25 years ago, Felsenfeld reported that "approximately 27 states...now prohibit the business of debt adjusting, that is debt adjusting for a profit," *id.* at 928–29, while "[t]here are some 14 states that regulate this business," *id.* at 929.

Felsenfeld had reason to know of the controversy surrounding the business of consumer debt adjustment. He had been appointed a consultant to the Commissioners on Uniform State Laws for the purpose of drafting a proposed article of the Uniform Consumer Credit Code to be entitled "Consumer Debt Counseling." *See id.* at 926. The preliminary draft of the article "took the position of outlawing the profit making pro-raters and permitting the business only by non-profit organizations, and such other organizations that may do it as a part of their other services. (Lawyers, for example, might engage in credit counseling.)" *Id.* at 931. Felsenfeld reports, however, that strong opposition from those who would lose their livelihood resulted in a decision "to reserve Article 7 [the proposed article on Consumer Debt Counseling] for future use after which there could be further study and, perhaps, a more deliberate decision." *Id.*

threatens financial stability, further deferment of payment will often serve only to compound the financial problems of debtors who seek it. For this reason, counselors must first provide a knowledgeable assessment of the debtor's need for an extension plan. If aid in budgeting alone is sufficient to restore the debtor's financial health, the counselor will provide that assistance only. Obtaining extensions from creditors in such a case may merely serve to cause the debtor to discount the financial resolve that is required of him. If the plan the debtor seeks also provides for forgiveness of late fees, a reduction in interest rates or both, it will also impose needless lost-opportunity costs on creditors.

While the need to restructure the debtor's payment schedules is a necessary requisite of a counselor sponsored plan, it is not the only one. The debtor must have a regular source of income that is sufficiently in excess of his living expenses to enable him to pay at least the principal amount of the obligations covered by the plan in a period that customarily does not exceed four years. This requirement has a qualitative as well as quantitative component, for the amount of income the debtor can commit to funding the plan is, to some extent, a function of his commitment to successful completion of a workout.

Credit counselors frequently gauge a debtor's commitment by requiring the debtor to propose where and in what amounts he will curtail recurring expenditures in order to fund the plan. When the debtor can defend his proposals against the counselor's devil's advocacy, the debtor signals his commitment to the credit counselor.[61] The four-year limitation that counseling agencies place on the term of the plan—increased from a three-year one in recent years—doubtlessly reflects the increased risk of the debtor's loss of commitment to a workout that extends over too long a period. Note too the parallel to payments under Chapter 13 plans in bankruptcy, which may not exceed three years unless the court for cause approves a longer period not to exceed five years.[62]

Although creditors' concern with the length of the plan is understandable, the reason for their imposition of the requirement that the plan provide for full payment of all the debtor's obligations except installment payments on long-term debt that come due after the term of the plan, such as payments on a home mortgage and, in some instances, the debtor's automobile, is not so readily apparent. Composition agreements, whereby cred-

---

61. Michael Spence defines "market signals" as "activities or attributes of individuals in a market which, by design or accident, alter the beliefs of, or convey information to, other individuals in the market." A. MICHAEL SPENCE, MARKET SIGNALING; INFORMATIONAL TRANSFER IN HIRING AND RELATED SCREENING PROCESSES 1 (1974). The author applies the signaling concept to the credit granting process, but not to collections. *See id.* at 69–75.

62. 11 U.S.C. § 1322(d) (1994).

itors agree among themselves to discharge a debtor upon his payment of some part, usually a pro-rata part, of each creditor's claim, have long received legal recognition. And while one creditor's sponsorship of a consumer's composition plan will probably encounter an even higher transaction-cost barrier than that creditor's sponsorship of an extension plan due to the need to obtain increased concessions from other creditors, the function of counseling agencies is to make transaction costs manageable. Why then are these agencies precluded from fashioning a plan that proposes less than full payment of claims in any instance in which that plan offers the promise of more recovery than its alternatives, coercive collection or bankruptcy? The same motive fuels a creditor's acceptance of both extension and composition plans: maximizing the salvage value of her claim.

The issue of proposed full payment of all claims is worth pursuing to understand the limits of creditors' trust in the work of counseling agencies and therefore the limits creditors place on their practice of not fully monitoring their monitor, the counseling agency. An understanding of these limits will in turn provide an informative insight on a question previously raised but not answered in this work: why do creditors embrace and financially support the work of counseling agencies but often question a debtor's invocation of Chapter 13 of the Bankruptcy Code?

Were a counselor given the discretion to propose something less than full payment of claims, any sympathy on her part for the debtor, which might be heightened by her close association with him, might motivate her to pursue acceptance of a plan that provided for less payment than the debtor could reasonably afford to make. In some cases, the credit counselor's incentive to propose such a plan might arise from a less noble but more tangible consideration—that of monetary compensation by the debtor. Because the costs of any such payments by the debtor to his counselor would ultimately be borne by his creditors, the counselor who decided to further her personal gain would be ideally situated to do so. The normal incentive of the debtor to curtail his expenses is obviously missing in this example of what economists describe as the problem of "moral hazard."[63] To limit transaction costs, the counselor alone gathers the relevant financial data on the debtor, and creditors are dependent upon both the counselor's acquisition of data and her sound judgment as to what may reasonably be expected of the debtor.

To contain this problem inherent in any agency relationship, creditors limit consumer credit counselors to plans that extend payment terms only. Although agency costs may also arise in extensions plans, for the coun-

---

63. See Mark V. Pauly, *The Economics of Moral Hazard: Comment*, 58 AM. ECON. REV. 531 (1968).

selor may propose a longer term for payment than necessary, the magnitude of such potential costs will normally be far less in those plans than they would be in counselor-sponsored composition plans.

By limiting the work of credit counselors to extension plans, creditors cast these counselors in a much more limited agency role than that of attorneys representing distressed debtors. The credit counselor's sole purpose is to accommodate the debtor's desire to pay his obligations in full. In performing that function, counselors may not assume an unrestricted role as advocate for either the debtor or his creditors. Their duty is neither to provide the debtor with the least costly relief from his debacle nor to provide creditors with the most expeditious collection of their claims, although both of these purposes will commonly be served to a considerable degree in a successful workout. To the extent that the debtor values full payment of his obligations more than he does relief from the claims of his creditors in other ways, such as discharge of his debts in bankruptcy, the benefits of a workout to the debtor will exceed its costs. And more expeditious payment of part of their claims will not benefit creditors whose total recovery is significantly reduced by failure to grant reasonable extensions.

Obviously, creditors could address the agency-cost problem that is present in composition plans in a manner that would not preclude credit counselors from sponsoring such plans. One or more of the creditors in a collection case in which the counselor proposed less than full payment of claims could confirm the data presented by the counselor and pass on the objectivity of her judgment. Assigning the role of monitoring the counselor to only one of the creditors avoids the costs of either duplicative efforts or undermonitoring among creditors. Undermonitoring occurs when every creditor seeks to reduce his own monitoring costs by "freeriding" on the efforts of others.[64] Neither undermonitoring nor duplicative efforts is, by definition, efficient.

But achieving optimal monitoring among creditors in this instance is not costless. Some creditor—the debtor's principal one is the most likely candidate for he has more at stake than others—must assume the costs of such monitoring, and other creditors must be convinced that he will safeguard their interests as well as his own. I cannot say that this never happens, but none of the counselors that I interviewed had tried to enlist the support of creditors in a composition plan. One credit counselor, with considerable experience in working with creditors in his community, did suggest that he was considering the possibility of proposing a composition plan when an ideal case came along.

Counselor-assisted workouts are not adapted to accommodate composition plans for the same reason that debtors frequently fail to obtain the sup-

---

64. *See* Levmore, *supra* note 40.

port of creditors for extension plans when the services of a credit counselor are missing. The transaction-cost problem again rears its ugly head. Once one or more creditors must monitor the work of the counselor, her work may be rendered largely redundant, and the cost savings she otherwise effects lost.

Another reason for precluding credit counselors from sponsoring composition plans probably exists. Creditors are loath to abandon the hope of collecting all their claims even in instances in which some significant loss appears virtually certain. Compositions force them to abandon that hope, regardless of how remote it may be. After all, the debtor just may have a rich uncle who just may intervene in his behalf. Like the late Justice Holmes, and probably many of us, creditors subscribe to the view that "certainty generally is illusion,"[65] and certainly this applies to certainty of loss as well as certainty of gain.

In certain respects, a creditor's abandonment of the hope of eventual collection of the unpaid portion of his claim is more justified when he is a party to a composition agreement than when his debtor receives a bankruptcy discharge. Most debtors will probably feel less compunction to pay a debt discharged by agreement than one discharged by order of a bankruptcy court. Moreover, courts give more favorable treatment to promises to pay remaining balances on dischargeable claims following bankruptcy than they do similar promises following compositions. The law has long recognized that a post-bankruptcy promise to pay a claim dischargeable in bankruptcy is enforceable because it is supported by a moral obligation,[66] although under modern bankruptcy law such reaffirmations must occur before the bankruptcy discharge is granted and must comply with various other safeguards designed to protect the debtor.[67] No "moral consideration" survives to support a new promise to pay following discharge of a debt in a composition agreement. Furthermore, such a promise to pay or actual payment may have disastrous effects for the debtor:

> Payment of a secret preference to one creditor, or the giving of preferential notes, or the promise of preferential payment, is generally held to entitle other creditors to rescind the composition agreement, whether done before execution of the composition agreement, in which case its concealment constitutes fraud, or after execution, in which event it constitutes a breach of the agreement.[68]

---

65. Oliver W. Holmes, *The Path of the Law*, 10 HARV. L. REV. 457, 466 (1897).
66. *See* Zavelo v. Reeves, 227 U.S. 625, 629 (1913).
67. *See* 11 U.S.C.A. § 524(c), (d) (West 1993 & Supp 1996).
68. VERN COUNTRYMAN, CASES AND MATERIALS ON DEBTOR AND CREDITOR 234 note 1 (1st ed. 1964) (citation omitted).

To understand why creditors are more skeptical of the protection given their interests in Chapter 13 proceedings than in workouts, contrast the credit counselor's role in the workout process, which is limited to the creation of extension plans, with the role of the debtor's attorney in Chapter 13 proceedings, which is not. In representing a debtor, the attorney's proper role is to explore with her client all lawful means of obtaining relief and to inform him of the costs of each alternative. Obviously, the costs of payments to creditors in a workout in which they are paid in full will exceed the costs of those payments in a bankruptcy action in which discharge is granted on less than full payment of claims.

To be sure, there are other costs that should be considered by the debtor in assessing the relative merits of workouts and bankruptcy, including the adjustment of debts in Chapter 13. Loss of ability to obtain credit in an economy that functions extensively on the use thereof is the principal economic concern of most debtors.

Credit counseling agencies sometimes emphasize this loss-of-credit cost of the bankruptcy alternative to workout in advertisements for their services. A striking example, due to its juxtaposition with another ad, is that run by Consumer Credit Counseling Services of Utah in the classified ads of a newspaper on a day in which I was passing through the airport in Salt Lake City. Appearing immediately below the ad of an attorney for his "DEBT REMOVAL CENTER," the counseling agency's ad is captioned BANKRUPTCY [:] A 10–YEAR MISTAKE." [69] The ad's forecast of 10 years of credit deprivation is no doubt based on a provision in federal law that prohibits a consumer reporting agency from making any consumer report—a credit report—of a bankruptcy case in which the order for relief antedates the report by more than 10 years.[70] Bankruptcy, however, may cast a longer than 10–year shadow on credit applications. A potential credit grantor may have knowledge of an applicant's prior bankruptcy from sources other than a current credit report and no provision in federal law, except a recent prohibition on denial of student loans made or guaranteed by a governmental agency,[71] prohibits the potential credit grantor from denying credit to an applicant based on a prior bankruptcy.[72]

The modern-day consumer who is without a source of credit, like the outlaw of earlier times who perforce retreated to the darker recesses of the forest, forgoes many of the amenities of organized society. Present grat-

---

69. THE SALT LAKE CITY TRIBUNE, Aug. 8, 1991 at B7.

70. *See* Consumer Credit Protection Act (Title VI. Fair Credit Reporting Act § 605, 15 U.S.C. § 1681(c) (1994).

71. *See* 11 U.S.C.A. § 525(c) (West Supp. 1996).

72. *See* Consumer Credit Protection Act (Title VII. Equal Credit Opportunity Act) § 701 15 U.S.C. § 1691 (1994).

ification and convenience are the most palpable functions of consumer credit. But that most common passport to these benefits, the credit card, often performs another function as well. Where merchants condition acceptance of a buyer's check on the presentation of a major credit card, a not uncommon practice, loss of credit standing affects the buyer's ability to make a common type of cash transaction. Nor are financial institutions eager to provide checking accounts and debit cards to individuals emerging from bankruptcy due to difficulties in ensuring that checking accounts are not overdrawn. While credit grantors and providers of other financial services do not relish an applicant's prior use of a workout, an applicant's prior bankruptcy normally inflicts far more serious harm on his ability to obtain credit and other financial services than a successfully completed workout.

I believe that at least the more sophisticated credit grantors do distinguish between workouts and bankruptcy in assessing an applicant's creditworthiness. While I have questioned only a limited number of creditors on this issue, certainly rational bases exist for the distinction. A debtor who has successfully completed a workout has exhibited a tenacity that may far surpass that required of a debtor in Chapter 13. And perhaps of even more importance to potential credit grantors, he has done so in an undertaking in which creditors were voluntary, even if reluctant participants and recovered at least the principal amount of their claims, except for certain reasonable collection costs, in full. Another factor enhances the chances of a debtor obtaining credit following successful completion of a workout. If his counselor feels he has learned from the experience—and successful completion of a difficult plan is good evidence of that—the counselor will frequently assist him in obtaining limited credit from retailers and lenders who do not limit their business to the most creditworthy sector of the market. In this manner, the formerly ostracized debtor may again become a card-carrying member of the credit society.

While one can make a theoretical case for the creditworthiness of former bankrupts, I do not believe that argument is well received by credit grantors, except perhaps those who operate in the high risk and most costly sectors of the market. Long ago and while more naive, I remarked, before an audience of credit union officials, that individuals recently discharged in bankruptcy might be good candidates for loans. I reasoned that credit grantors would face less competition with other creditors for payment and the debtor would be interested in faithfully performing his obligation to restore his credit rating. My theory, and it was nothing more than that, was not well received. In fact, it was soundly booed. Discussions with representatives of creditors since that time have convinced me that my audience did not react aberrantly. Even if a former bankrupt, aside from that bankruptcy, presents a good case for a credit extension, I believe that credit grantors are prone to deny his application. Is this action irrational for a creditor who needs to increase his loan volume? Perhaps, but it is also possible that that creditor justifies his action on the basis that it sends a clear message to other debtors who contemplate bankruptcy. In recent years, however, there has been, judging by advertising in the media and other evidence, a growth in the suppliers of credit to substandard borrowers. Witness the ads of used-car dealers that pitch their advertisements to former bankrupts or other people with bad credit, but note that this type of lender typically charges much higher rates and also often lends on restrictive terms, such as requiring that the debtor maintain a security deposit with the credit-card issuing bank for a so-called "secured credit card."[73]

The stigma of bankruptcy may harm aspects of the debtor's reputation in addition to his credit rating. It may adversely affect his employment opportunities,[74] aspects of his business reputation in addition to that of creditworthiness, his civic and community standing, and his social status. These costs are not as commonly associated with successful workouts, and a debtor's perseverance in the face of economic adversity may even benefit his reputation in one or more of these respects.

Harry Truman's resolve to settle his debts outside of bankruptcy is a notable case in point. A partner in a haberdashery store that failed in the adverse economic climate of the early 1920's, Truman refused the advice of lawyers to file a bankruptcy petition when his partner did so. Instead, he insisted upon paying off all the debts the partnership had incurred. Eventually, Truman settled the claim for the unexpired term of the store's lease and purchased through his brother a note from the receiver of a failed bank that held it. Although his efforts resulted in less than full payment of these claims, Truman struggled for years with the debts he had incurred in connection with the haberdashery business.[75]

Excising debt in bankruptcy also imposes a psychic cost, at least on some debtors, that payment of obligations in workouts does not. To avoid the loss of self-esteem that they associate with "taking bankruptcy," some

---

73. *See* ELIZABETH A. WARREN & JAY LAWRENCE WESTBROOK, THE LAW OF DEBTORS AND CREDITORS 433 (3d ed. 1996).

74. Neither a governmental unit nor a private employer may discriminate with respect to employment against an individual "solely because" of bankruptcy. *See* 11 U.S.C.A. § 525 (West 1993 & Supp. 1996).

75. ALFRED STEINBERG, THE MAN FROM MISSOURI 57–58 (1962); *see also* JONATHAN DANIELS, THE MAN OF INDEPENDENCE 108 (1950) ("Across the years, he settled...claims on the basis of the best settlement he could get, but they hung heavy on him even after he was Senator, more than a decade later.").

individuals will labor mightily to pay their obligations, even in instances in which the source of overextension was non-volitional, such as job loss or expenses incurred for health care. For them, adherence to the Emersonian model of self-reliance[76] in a contemporary setting—paying his or her own way in an increasingly interdependent world—is singularly worth the costs a workout entails.

Notwithstanding these costs of a bankruptcy discharge, many debtors who seek legal help decide that some form of bankruptcy affords them greater net gains than do workouts. Of course, debtors who initially contact lawyers instead of credit counselors will often know, before any consultation with a lawyer, that the lawyer draws upon a wider range of remedies than the counselor. Therefore these debtors may constitute a self-selected group of individuals who are inclined to choose bankruptcy over workouts before a lawyer has had any opportunity to influence their choice. The decision, however, of many debtors to use a lawyer and bankruptcy in preference to a credit counselor and a workout may be based on something other than a predisposition to choose one viable alternative over another. Often the debtor consults a lawyer because payment of all her obligations in a workout is simply beyond the scope of any reasonable test of her ability. While authorities differ on the issue of whether a significant number of debtors filing Chapter 7 cases—which often result in little or no payment to unsecured creditors—could have paid a considerable amount of their debts in a Chapter 13 action, certainly fewer of those Chapter 7 filers could have met the higher standard of full payment that creditors impose on counselor-assisted workout plans.

Many creditors, however, advance another reason for a debtor's choice of bankruptcy over workouts, which also serves as an indictment of some members of the bankruptcy bar. They charge these lawyers with failure to advise their clients of the loss of ability to obtain credit and other harmful aspects of bankruptcy. In more serious form, their complaint is that these lawyers work to balance the scales in favor of bankruptcy even when the debtor is aware of these attendant costs. When the lawyer is neither constrained by countervailing ethics nor an interest in long-term representation of his client, there is some basis for their charges. Lawyers earn fees from filing bankruptcy actions, not from referring clients to credit counseling agencies, and studies show that bankruptcy is the remedy of choice for debtors after they consult attorneys, even when other alternatives that seem preferable exist.[77]

---

76. *See* RALPH W. EMERSON, *Self Reliance*, *in* ESSAYS: FIRST SERIES (1841), *reprinted in* THE COMPLETE WRITINGS OF RALPH WALDO EMERSON 138 (1929).

77. *See* Jean Braucher, *Lawyers and Consumer Bankruptcy: One Code, Many Cultures*, 67 AM. BANKR. L.J. 501 (1993); Gary Neustadder, *When Lawyer and Client Meet:*

While lawyers might earn their fees by representing consumer debtors in workouts, like they frequently do in cases of overextended commercial debtors, the economics of consumer cases does not make this type of representation attractive to many of them. The debtor's disposable income is often scarcely enough to fund an extension plan, lawyers charge more than credit counselors and the costs they incur to obtain creditors' support for the workout are generally higher than the costs incurred by credit counselors, who benefit from economies of scale in their dealings with creditors. To appeal to creditors, a lawyer may agree to collect legal fees and costs as or after creditors are paid. With this method, however, the lawyer incurs the risk of non-recovery of his claim and, no doubt, the skepticism of creditors that he will take that risk. If he insists on a retainer to cover his higher fees and costs, that retainer takes priority over creditors' claims and considerably diminishes the probability that creditors will acquiesce in the workout. At best, the lawyer must exert herculean effort compared to that of the credit counselor to obtain the support of creditors for such a plan.

By the foregoing analysis, however, I do not mean to deny any role to lawyers in optimally resolving consumer-debt problems in a non-bankruptcy setting. As a lawyer may credibly bolster her offer for a plan that necessarily proposes a reduction in principal as well as extensions in payment by mention of the proposed alternative of a more costly bankruptcy filing, the lawyer's unique strength lies in her ability to custom tailor the debtor's proposed plan to the debtor's financial means for the benefit of all parties.

Because the debtor's lawyer, however, is often faced with the two unacceptable alternatives of deferring fees or alienating creditors by taking a retainer for work to be done in implementing workouts, he will tend to favor a bankruptcy action in any instance in which the invocation of that remedy is justifiable, even if only marginally so. In that setting, he is not faced with the daunting task of convincing skeptical creditors, who may demean his services, of the merits of the remedy he proposes for the debtor. Instead of playing the subservient role of importuner of favors from creditors' hard-nosed collectors, the attorney casts himself in a commanding position and performs services that only lawyers can provide. That these services should command fees commensurate with the particular skills of his profession is, of course, a given.

Most consumers probably attach a higher value to representation of their interests in judicial proceedings than in out-of-court negotiations. Besides, in a bankruptcy action, the lawyer's fee may be paid from property that would otherwise be surrendered to the trustee for payment to

---

Observations of Interviewing and Counseling Behavior in the Consumer Bankruptcy Law Office, 35 BUFF. L. REV. (1986).

creditors. This fact may greatly reduce the debtor's incentive to monitor the reasonableness of that fee. A provision in the Bankruptcy Code recognizes this problem of moral hazard, and limits an attorney's fee "for services rendered or to be rendered in contemplation of or in connection with" a bankruptcy case to the "reasonable value" of the attorney's services.[78]

The factors that cause some debtors and the lawyers advising them to favor bankruptcy relief, even when the debtor would face no undue hardship by full payment of his obligations under an extension plan, need not produce less than complete recovery of creditors' claims. Bankruptcy law protects the interests of creditors in such instances in two primary ways.

First, the gates to Chapter 7 of the Bankruptcy Code are no longer open to just any debtor who comes calling. Creditors are probably most offended by those Chapter 7 cases in which a debtor with a substantial income receives a discharge upon surrendering nonexempt assets of little value to his creditors.[79] As creditors count primarily on a consumer debtor's regular income and not liquidation of assets of little market value, creditors and some theorists have argued that that debtor should be required to commit at least some of his future income to the payment of creditors as the price of his discharge.[80] That principle is now reflected in a provision of Chapter 7 of the Bankruptcy Code that empowers the bankruptcy court "on its own motion or on a motion by the United States Trustee" to dismiss a voluntary Chapter 7 case filed by an individual "whose debts are primarily consumer debts" upon a finding of "substantial abuse" of the provisions of Chapter 7.[81] While the vague statutory standard of "substantial abuse" does not expressly incorporate a test based on the ability of the debtor to make substantial payments from future income, significant judicial decisions applying the standard have imposed that test.[82] But the standard is admittedly vague, and survey results reveal significant dispar-

---

78. 11 U.S.C. § 329 (1994).

79. *See* ROBERT L. JORDAN & WILLIAM D. WARREN, BANKRUPTCY 611 (3rd ed. 1993).

80. *See* Theodore Eisenberg, *Bankruptcy Law in Perspective*, 28 UCLA L. REV. 953, 980 (1981) ("Repayment plans that draw upon future earnings more accurately reflect a debtor's ability to pay than do liquidation plans. We live to a great extent in a cash flow world.").

81. *See* 11 U.S.C. § 707(b) (1994).

82. *See In re* Walton, 866 F.2d 981 (8th Cir. 1989); Zolg v. Kelly (*In re* Kelly), 841 F.2d 908 (9th Cir. 1988). While these cases hold that a debtor's income in excess of expenses, standing alone, constitutes a substantial abuse of Chapter 7 justifying dismissal, *Green v. Staples* (*In re Green*), 934 F.2d. 568 (4th Cir. 1991) holds that the debtor's ability to repay is the primary but not sole factor to be considered in applying the "substantial abuse" test.

ity in the frequency of use of substantial-abuse motions by various bankruptcy judges and U.S. Trustees.[83]

When a debtor does lose access to Chapter 7, however, she is not precluded from filing a Chapter 13 petition. But bankruptcy law now provides, at least in theory in Chapter 13 if not always in fact, better protection of unsecured creditors, the class that suffers worse in any bankruptcy, than they would receive in Chapter 7. While unsecured creditors as a class do not get to vote on confirmation of a plan in Chapter 13 as they do in a business reorganization in Chapter 11, they are protected by the requirement that the plan must be proposed in "good faith" and, further, by the "best interests" test that requires that each creditor receive at least as much as she would have received in a Chapter 7 bankruptcy of the debtor.[84] Of greater significance, however, since 1984 Chapter 13 has provided that any unsecured creditor may block confirmation of a plan that provides for less than full payment of her claim, unless "the plan provides that all of the debtor's projected disposable income to be received in the three-year period beginning on the date that the first payment is due under the plan will be applied to make payments under the plan."[85] For the purpose of applying this rule, the term, "disposable income," means, in the context of a consumer debtor, "income which is received by the debtor and which is not reasonably necessary to be expended for the maintenance or support of the debtor or a dependent of the debtor."[86]

Thus, it may appear that even unsecured creditors in Chapter 13 cases, while not necessarily being promised payment in full, receive in one important sense the equivalent of what they get in counselor-assisted workouts. Arguably, creditors in both instances get the best the debtor can do in a long period—three-years in Chapter 13 and four-years in workouts—of curtailing expenses in order to pay obligations. Yet, if the protection of creditor interests is essentially equal, my conclusion that creditors are often disposed to view Chapter 13 actions as less protective of their interests

---

83. Wayne R. Wells et al., *The Implementation of Bankruptcy Code Section 707 (b): The Law and the Reality*, 39 Cleve. St. L. Rev. 15, 42 (1991). Meanwhile, the debate over a needs-based bankruptcy system and congressional formulation of that system is fueld by proposals for further legislation, heated debates by the National Bankruptcy Review Commission, and continuing studies by private agencies. *See* 9 Bankr. L. Rev. 310–11, 318 (Oct. 23, 1997).

84. 11 U.S.C. § 1325 (a)(3), (4) (1994).

85. *Id.* § 1325(b)(1) (1994). To the extent that a creditor holds a secured claim, *see id.* § 506(a), he is protected in any case in which the debtor retains his collateral by the requirement that the plan provide for retention of his lien and for payments, having a value as of the effective date of the plan, of the amount of his secured claim. *See id.* § 1325(a)(5).

86. *Id.* § 1325(b)(2)(A).

than counselor-assisted workouts is either erroneous or inexplicable. There are, however, differences between the two methods that explain the creditors' preference for the counselor-assisted workout.

The two types of collective undertakings first vary in the degree of commitment a debtor must bring to each. Under the bankruptcy standard, what is necessary for the support of the debtor and his dependents is no doubt a flexible standard. While there are doubtlessly some bankruptcy judges who apply that test in a manner that requires the debtor's maximum commitment to payment of his debts in a Chapter 13 plan, there are probably a large number of others who apply a less stringent standard. One bankruptcy judge who conversed with me on the subject of credit-counseling agencies concluded unequivocally that the agencies were not needed in his district as the Chapter 13 plans that he confirmed were fully protective of creditors' claims. There are probably wide differences in the standards that are imposed on Chapter 13 plans by the various bankruptcy judges and consequently like divergences in creditors' views of the efficacy of Chapter 13 actions.[87]

In comparison, a debtor in a workout whose objective is full payment of his claims must be far more apt to curtail his living expenses and perhaps moonlight to increase his income than a Chapter 13 debtor. The latter need only meet a test of the reasonableness of his living expenses and is subject to that lesser standard for a period that may be one year less than he would encounter in a workout. Given the greater incentive of a debtor in a workout to effect full payment, for unlike Chapter 13 proceedings his debts are discharged only if he does so,[88] creditors justifiably perceive workouts, where feasible, as preferable to Chapter 13 plans.

Other reasons justify the distinction creditors make in workouts and proceedings in Chapter 13. The statute requiring that all the debtor's disposable income be directed to payments under the Chapter 13 plan[89] may produce little for unsecured creditors if the debtor directs that income in large part to payments on secured claims.

That statute does not restrict the debtor from doing so, and another provision of the Bankruptcy Code permits Chapter 13 plans to "provide for the curing of any default within a reasonable time and maintenance of payments while the case is pending on any unsecured claim or secured

---

87. *Compare* 5 LAWRENCE KING, COLLIER ON BANKRUPTCY ¶ 1325.08 [4] [b] (15th ed. 1992) (concluding that a judge's determination of the debtor's disposable income should not mandate drastic changes in the debtor's lifestyle) *with In re* Kitson, 65 B.R. 615, 622 (Bankr. E.D. N.C. 1986) ("A Chapter 13 debtor who proposes to pay his creditors 38 cents on the dollar cannot expect to 'go first class' when 'coach' is available.").

88. See 11 U.S.C.A. § 1328 (West 1993 & Supp. 1996).

89. *See* 11 U.S.C. § 1325(b) (1994).

claim on which the last payment is due after the date on which the final payment under the plan is due."[90] Chapter 13 encourages debtors to channel their contributions to the plan toward payments on these long-term debts, which are usually secured by the debtor's home, because Chapter 13 excepts such long-term debts from its discharge.[91] Secured creditors whose claims are payable before the expiration of the term of the Chapter 13 plan will not have their debts excepted from discharge. When the debtor retains their collateral, however, they may exact a promise of payments having a value, as of the effective day of the plan, of the amount of their secured claims.[92]

Often, the debtor's need or desire to retain his home, automobile, boat or other encumbered property will dictate that his plan provide for payments to secured creditors at the expense of unsecured ones. The result of these provisions is that in spite of the disposable income test, a Chapter 13 plan may be confirmed that provides for no payment to unsecured creditors.[93] A statistical study showing that Chapter 13 debtors had considerably more secured and less unsecured debt than debtors in Chapter 7[94] evinces that some consumers may be attracted to Chapter 13 not to increase the recovery of their unsecured creditors but to save encumbered property that they would lose in Chapter 7.

Unsecured creditors fall far short of full recovery of their claims in Chapter 13 despite the disposable income test. Researchers have found that average proposed payments range from about one-third to one-half of unsecured claims. Moreover, in only about one third of the Chapter 13 cases do debtors complete payments of what was promised in the plan.[95] Workouts sponsored by consumer credit counseling agencies propose full payment to creditors and have a higher success rate. The National Foundation for Consumer Credit reports that 25% of their debt management plans are completed and another 22% become self-administering when

---

90. *Id.* § 1322(b)(5).
91. *See* 11 U.S.C. § 1328(a)(1),(c)(1) (1994).
92. *See* 11 U.S.C. § 1325(a)(5) (1994).
93. *See In re* Greer, 60 B.R. 547 (Bankr. C.D. Cal. 1986) (husband and wife co-debtors had a sizeable arrearage to cure on their home mortgage and chose to retain their two automobiles by making installment payments on their debts secured by the vehicles). But the Bankruptcy Code also requires that Chapter 13 plans be proposed in "good faith," *see* 11 U.S.C. § 1325(a)(3) (1994), and some bankruptcy judges have denied confirmation of plans providing for no payments to unsecured creditors on that basis. *See, e.g., In re* Lattimore, 69 B.R. 622 (Bankr. E.D. Tenn. 1987).
94. REPORT OF THE COMPTROLLER GENERAL TO THE CHAIRMAN OF THE COMMITTEE ON THE JUDICIARY, HOUSE OF REPRESENTATIVES 47 (1983).
95. *See* WARREN & WESTBROOK, *supra* note 73, at 438.

there is felt to be no further need for the use of the counseling agency as intermediary between the debtor and her creditors. Counseling agencies close debt management plans for failure of the debtor to perform in 48% of their cases, and debtors leave counseling agencies for bankruptcy courts in the remaining 5% of debt management plans.[96]

In counselor-assisted workouts, a secured creditor does not normally extend terms of payment, for doing so, while depreciation of collateral goes unchecked, would diminish the value of the collateral relative to the amount of the claim. In this manner, creditors participating in workouts recognize the bankruptcy principle of adequate protection of secured claims.[97] But in counselor-assisted workouts, the cost of this practice to unsecured creditors is limited by the requirement that the plan provide for payment of their claims in full. While unsecured creditors in these workouts suffer further delays to accommodate the debtor's needs to bring and keep payments on secured claims current, the promise of ultimate payment on their claims may not be a casualty of those needs.

Unsecured creditors also prefer a workout to bankruptcy because they will benefit from the absence of an attorney's fee and court costs. These costs are borne by creditors except where they are paid by a friend or relative of the debtor or the debtor uses property that would not have funded a Chapter 13 plan,[98] been distributed to creditors in a Chapter 7,[99] or used for prepetition payment to creditors. Even though credit counseling agencies generally succeed in their efforts to get creditors for whom they collect to defray their costs of operation—a charge of some 12 percent on collections is commonly touted as the creditor's "fair share"—workouts will commonly entail lower costs than attorney's fees and court costs in a bankruptcy proceeding.[100]

---

96. The success rates of consumer credit counseling agencies in debt management plans were provided me in November 1996 by Durant S. Abernethy, president of the National Foundation for Consumer Credit.

97. *See* 11 U.S.C. § 361 (1994).

98. Chapter 13 plans are generally funded by the debtor's salary or wages earned after the filing of the bankruptcy petition. The Bankruptcy Code's general provisions define property of the estate primarily in terms of the debtor's prepetition property. *See* 11 U.S.C.A. § 541 (West 1993 & Supp. 1996). To accomplish its goal, Chapter 13 supplements these general provisions by including the debtor's postpetition property including earnings as property of the estate. The debtor remains in possession of all property of the estate, however, except that used to fund the plan. *See* 11 U.S.C. § 1306 (1994).

99. In a Chapter 7 proceeding, the debtor will retain his exempt property. *See* 11 U.S.C.A. § 522 (West 1993 & Supp. 1996).

100. Large creditors in a community will also support their local credit counseling agency by donations of money and in-kind contributions such as office space or equipment. While this support increases the cost to the donors of counselor-assisted workouts,

A final reason that unsecured creditors prefer counselor-assisted workouts to Chapter 13 actions is based not on differences in the recovery of principal but on a principle that creditors share: participation in debt adjustment by consent is generally preferable to participation by command of government. In the closest parallel to debt adjustment in Chapter 13, a business reorganization in Chapter 11, unsecured creditors as a class do have a say in approving the plan.[101] But there is no similar provision in Chapter 13 giving unsecured creditors, even by class, a negotiating stance in the formulation of the plan.[102] Chapter 13 does provide that the court may not confirm a plan unless it provides for payments to each unsecured creditor that have a present value not less than the amount that creditor would have received if the estate of the debtor had been liquidated in Chapter 7.[103] This provision, however, gives unsecured creditors little bargaining strength as Chapter 7 bankruptcies by consumer debtors often produce little or nothing for unsecured creditors.

In contrasting the treatment of creditors in counselor-assisted workouts with their treatment in Chapter 13 proceedings, attention must also be given to secured creditors in the two proceedings. While debtors may channel the disposable income they are required to commit to their Chapter 13 plans to their secured creditors in order to retain their encumbered property, it does not follow that secured creditors do better or even as well in a Chapter 13 proceeding as they do in a counselor-assisted workout.

While a secured creditor is entitled to payments under the Chapter 13 plan in the amount of his secured claim plus interest over the life of the plan,[104] the amount of the secured claim, a defined term in bankruptcy, cannot exceed the value of the collateral.[105] Hence, where the value of the collateral, as determined by the bankruptcy court, is less than the amount of the creditor's claim, the debtor may retain the collateral by proposing amortization of the lesser amount plus interest thereon in payments under the plan.

A restriction on this right of the Chapter 13 debtor to modify secured claims protects the mortgagee of the debtor's home from this "lien strip-

---

it also testifies to the value they believe they receive from the work of the counseling agency.

101. *See* 11 U.S.C. § 1129(a)(8) (1994). While a Chapter 11 plan may be confirmed over the objection of a class of unsecured creditors by resort to "cramdown," *see id.* § 1129(b)(2), the proponent of the plan will typically find it advantageous to seek confirmation by negotiation. *See* Richard F. Broude, *Cramdown and Chapter 11 of the Bankruptcy Code: The Settlement Imperative*, 39 Bus. Law. 441 (1984).

102. 11 U.S.C. § 1325 (1994).

103. See id. § 1325(a)(4).

104. 11 U.S.C. § 1325 (a)(5)(B)(ii) (1994).

105. 11 U.S.C. § 506(a) (1994).

ping."¹⁰⁶ But security interests in motor vehicles, boats, furniture, appliances and other personal property may be reduced to the court determined value of the collateral and the creditor protected as a secured creditor only to the extent of that value. The balance of the creditor's claim is treated as an unsecured one. The partial stripping of a creditor's lien in Chapter 13 is a more effective device for debtors than a similar provision limited to certain tangible personal property in Chapter 7, because the latter provision requires the debtor make a lump-sum payment of the value of the collateral to redeem it.¹⁰⁷ In contrast to the treatment of secured creditors in bankruptcy, in counselor-assisted workouts secured creditors may insist on their rights under state law to receive full payment of their claims before loss of their security interests.

The foregoing analysis of why creditors prefer counselor-assisted workouts to Chapter 13 proceedings does not mean, however, that creditors reject any meaningful role for that bankruptcy proceeding in protecting their interests. To the contrary, the consumer-credit industry has worked diligently to promote the use of Chapter 13 over Chapter 7, and its lobby was influential in imposing the "substantial-abuse" test previously examined on access to Chapter 7.¹⁰⁸ To better understand the function of counselor-assisted workouts in the collection of consumer debts and why creditors prefer it to either form of bankruptcy, it is necessary to understand the bases for and the extent of creditors' support for Chapter 13 over Chapter 7.

In some cases, creditors will have no reason for preferring Chapter 13 to Chapter 7. As previously noted, a Chapter 13 may produce no better results for unsecured creditors than a Chapter 7. Where creditors in such a case believe that they could have done better had bankruptcy not been invoked, "a plague a'both your houses"¹⁰⁹ will likely be their response to both forms of relief for consumers in the Bankruptcy Code.

In cases where creditors do support the use of Chapter 13, they may do so simply because they perceive it as less harmful to their interests than Chapter 7. These are cases in which a Chapter 13 will produce more for creditors than Chapter 7, but less than their probable recovery if bankruptcy had not been invoked.

Chapter 13 is predicated on something more, however, than being less disadvantageous to creditors than Chapter 7. In this additional function, it affords creditors benefits that are obtainable in no other way. Where

---

106. 11 U.S.C. § 1322(b)(2) (1994); Nobelman v. American Savings Bank, 508 U.S. 324, 113 S.Ct. 2106, 124 L.Ed2d 228 (1993).
107. 11 U.S.C. § 722; In re Bell, 700 F.2d 1053 (6th Cir. 1983).
108. *See* JORDAN & WARREN, *supra* note 79, at 611–13.
109. WILLIAM SHAKESPEARE, ROMEO AND JULIET, Act 3, scene 1, line 94.

full payment of claims in a counselor-assisted workout is not feasible and a Chapter 13 plan holds forth the promise of greater recovery than Chapter 7 or the running of the creditors' race of diligence, Chapter 13 will be seen by creditors as providing a positive benefit. There may also be cases in which creditors are indifferent to a debtor's choice of a counselor-assisted workout or a Chapter 13 proceeding. These cases will occur when the Chapter 13 plan provides for the same full payment of claims as the workout, provided the debtor is as committed to the Chapter 13 as to the workout. Very few Chapter 13 plans are limited to extensions, however, and most promise something considerably less than full payment of claims.

Because creditors realize that Chapter 13 is the best device for protecting their interests where collective action is necessary and counselor-assisted workouts are infeasible, they recognize an important role for Chapter 13 in the collection process. Obviously, Chapter 13 also benefits debtors who wish to pay more on their obligations than they would in a Chapter 7 or without bankruptcy assistance and effectively use Chapter 13 to do so. Walter Chandler, the sponsor of the 1938 amendment to the former Bankruptcy Act that enacted the forerunner of the present Bankruptcy Code's Chapter 13, observed that that legislation gave the debtor an opportunity "to look his creditors in the face instead of crossing to the other side of the street in order to avoid them."[110]

The important role that creditors recognize for Chapter 13 does not detract, however, from the role they assign the credit counselor. Creditors perceive a hierarchy of need by troubled debtors, extending from those who may still make full payments with the assistance of a counseling agency, through those who require a scaledown of their debts in a Chapter 13, to those for whom even serious effort at meaningful payment from future income will be fruitless, the proper candidates in creditors' estimation for Chapter 7 bankruptcies.

The best evidence that credit counseling agencies play an increasingly important role in providing a suitable remedy when debtors are at one level of this hierarchy of need—a remedy that furthers the interests of debtors who wish to pay their obligations as well as the interests of their creditors in receiving that payment—is found in the dramatic growth of these agencies in recent years. From humble beginnings in the mid-1950s, there are now better than 1200 consumer credit counseling offices in the United States and Canada.

In stating the case for credit-counseling agencies, I have focused on the role they play in a collection process that otherwise provides no systematic means to obtain collective action when bankruptcy may still be avoided. Systematic solutions are particularly needed in consumer-collection cases

---

110. Walter Chandler, *The Wage Earner's Plan: Its Purpose*, 15 Vand. L. Rev. 169, 170 (1961).

because these solutions make transaction costs, which must be tightly contained in those cases, manageable. But like off-the-rack suits, the use of a credit-counseling agency does not always provide a good fit with creditors' and debtors' needs.

For example, creditors may be harmed when a credit counselor procures extensions that do not produce increased recovery of their claims but are used by the debtor to forestall the individual collection efforts creditors would otherwise make. This type of counselor's error may also increase the costs of the counselor's services, and thus the cost of the collection process, in subsequent cases. Too many disappointing recoveries will undermine a creditor's confidence in the counselor's judgment. This may result in the counselor encountering greater difficulty in getting that creditor to forego coercive collection measures in later cases when the implementation of workout plans is clearly justified.

Nevertheless, the growth in credit counseling evinces that this is not a serious problem. In large numbers, creditors, especially experienced ones, do accept the work of counseling agencies and will assist those agencies in securing the cooperation of other creditors who initially fail to acquiesce in workout plans. Not all commentators, however, sound so optimistic a note on creditors' cooperation in counselor-assisted workouts. Robert Jordan and William Warren, law professors at the University of California, Los Angeles and experienced observers of debtor-creditor law and practice, note that although "[n]onprofit credit counseling agencies are present in most large cities..., [n]o satisfactory private solution has been found that will enable a debtor who wishes to pay debts on a restructured basis to do so with protection against recalcitrant creditors."[111]

The remedy that counseling agencies offer may also fail in some instances to meet the needs of debtors who use it. A consumer who opts for a counselor-assisted workout without properly assessing the difficulties that the workout will entail may be better advised to file a Chapter 13 or Chapter 7 bankruptcy action. I have talked with bankruptcy attorneys and one prolific commentator on bankruptcy law who believe that many consumers pay too high a price for counselor-assisted workouts, that these agencies that are supported by creditors are not sensitive enough to the needs of debtors. Voicing this concern with he who pays the fiddler calling the tune, the Federal Trade Commission in March 1997 announced that with its guidance the National Foundation for Consumer Credit has established a new policy to insure that debtors seeking help from member consumer credit counseling agencies be informed that these agencies receive most of their funding from creditors.[112]

---

111. JORDAN & WARREN, supra note 79, at 610.
112. *See Credit Counseling Agency to Declare Relationship With Creditors, FTC Says,* 9 BANKR. L. REV., (BNA) 355 (Mar. 20, 1997).

The complaint of those who view credit counseling agencies as being overly zealous in protecting creditors is the converse of that of creditors who believe that attorneys channel many consumers into bankruptcy when workouts would be feasible and advantageous to the debtor as well as her creditors. Doubtlessly, there is some truth in the charges of both groups, and turf battles are bound to arise in those cases in which a good argument can be made for either form of relief.

I have been told by some credit counselors that their agencies do not permit them to recommend the services of a bankruptcy attorney to debtors, probably because the agencies feel that such action would undermine their relations with creditors upon which their financial support and ability to intercede on the part of debtors depend. To the extent this limitation exists, it is probably of little significance. Implicit in the counselor's failure to initiate a workout plan is the message that the debtor needs greater latitude in dealing with creditors than mere extensions on the terms of her obligations. And all debtors exposed to media advertising are aware of the additional remedies that bankruptcy courts and lawyers who access them can bestow.

There are doubtlessly instances in which a counselor does sponsor a plan that a better informed debtor would not have undertaken. Such plans will no doubt often be aborted, and this may impose additional costs on both the debtor and his creditors. Moreover, even when these plans are fully performed, they may impose costs on the debtor that transcend all benefits the debtor derives.

One such cost of the use of credit-counseling agencies to some debtors is a psychic one that results from having to forego the sole management of their financial affairs. This factor is probably most acutely felt by prudent, well educated people whose overextensions are the result of non-volitional expenditures or sudden losses of income and not mismanagement of debt. To avoid this humbling experience, some people will forego the use of a counseling agency and attempt to personally obtain their creditors' support for a workout. While no doubt many are ultimately successful, their task is no doubt made far more difficult by their failure to invoke the assistance of a credit counseling agency.

The costs to a debtor of entering into a counselor-assisted workout will exceed its benefits only when the debtor significantly underestimates the hardship the workout will entail. If the debtor's estimate of this hardship is more accurate, the fact that the debtor undertakes the workout when many other people similarly situated would choose bankruptcy is of no moment. The debtor is best able to assess his relative costs. Tastes for difficult workouts are entitled to the same respect as tastes for boiled okra or sushi.

It is all too easy to exaggerate the imperfections in the counselor-assisted workout method of systematic debt adjustment. Notwithstanding imperfections, credit counseling works better than any other method in cases that fall within its domain. As noted above, the remedy may cause harm to a debtor, her creditors, or both when it is applied to a case in which it affords no clear gain to one or both of them. The credit-counseling system has, however, a significant safeguard built into it to prevent its misapplication. Counselors are sufficiently motivated to objectively assess the probability of a successful workout and to decline those cases where the probability seems remote. They lose credibility with the creditors and debtors they serve, and perhaps their jobs, when they sponsor too many unsuccessful cases.

From the perspective of creditors, a case is not necessarily unsuccessful if the debtor fails to make full payment of his obligations under the plan. If the counseling plan recovers an amount greater than the probable recovery from any other manner of collection, the failed plan is still a relative success for creditors. As reflected earlier in the payoff matrix for the Creditor's Dilemma, the probable benefits that justify concerted action by creditors in workouts will never accurately reflect full payment of claims, for no means of collection is riskless.

One has more difficulty, however, in finding benefit to a debtor in a plan that is not completed, and here critics of counselor-assisted workouts may have cause for concern. A debtor who must seek relief in bankruptcy following an unsuccessful attempt at a workout has probably not enhanced his credit rating by the failed workout. Except to the extent that the debtor values his efforts to avoid bankruptcy in excess of their costs, he will have obtained no benefit from those efforts.

Perhaps the best answer to this criticism of counselor-assisted workouts is that any opportunity entails risks that should be carefully weighed against potential benefits. The debtor makes the final decision to undertake the workout. Credit counselors in nonprofit agencies have no financial or other incentive to encourage plans that are probably beyond the debtor's abilities. A credit counselor may be motivated to sponsor less than promising cases if his job is threatened by a declining case load. This problem would appear to be more theoretical than real, however, so long as counselors, like bankruptcy lawyers, are employed in a growth industry. As a counselor's job is made considerably harder by sponsoring a plan that probably exceeds the debtor's abilities, any deviation by the counselor from a purely objective assessment of the debtor's probability of successfully completing the plan is likely to be one that prejudices the counselor against taking the case, not one that commits him to encouraging a debtor's quixotic endeavor.

6

The Significance of the Systematic Resolution of the Creditor's Dilemma

A. Preserving Creditors' Remedies While Restricting Their Counterproductive Applications

The foregoing analysis of the role of credit-counseling agencies in collection practices in this country reveals two significant facts that should be of interest to lawmakers and theorists. First, creditors have successfully designed a private-sector device to supplement the public one of bankruptcy for systematically surmounting the transaction-cost barrier to beneficial collective action in cases of Creditor's Dilemma. The fashioning of this private-sector solution to the problem of overly coercive collection of consumer debt would warrant far less attention from those concerned with debtor-creditor law were it not coupled with the second of these facts. Rapid growth in the number of counseling agencies and the widespread distribution of these agencies throughout many areas of the United States have now made their services readily available to a large number of overextended consumers. In both theory and practice then, many debtors who are faced with debilitative dunning, wage garnishments or levies on their tangible property, and who do not wish nor need to resort to Chapter 7 or Chapter 13 bankruptcy, carry the key to initiating their economic rehabilitation: they need merely enlist the services of their local consumer credit counseling office.

Of what significance to lawmakers is this ability of the debtor to cage the creditors' hounds before the hunt begins in earnest and irreparable harm is done the hunters and the hunted? I believe that the emergence of a systematic method of avoiding harmful collection practices should have a significant impact on the limits lawmakers place on creditors' remedies in cases in which the debtor neither pays nor resorts to a counselor-assisted workout or bankruptcy.

Those who would abolish entirely some of the remedies traditionally used in the "race of diligence" by creditors must prevail in an important argument. They must establish that abolishment is the only generally effective means of prohibiting the use of those remedies in instances in which they harm the debtor or other creditors more than they benefit the creditor who employs them. But when the function of credit counseling agencies is recognized as a supplement to the two forms of bankruptcy relief, effective means exist for preventing the harmful use of creditors' remedies without abolishing their use in cases where they are justified.

No serious effort has formerly been made to normatively evaluate consumer collection law and practice from a perspective that recognizes the emergence of counseling agencies as major participants in the collection of consumer debt. But no attempt to set the boundaries of permissible creditors' remedies and collection practices should fail to address the impact of the work of these agencies in thwarting counterproductive coercive collection. Before examining how recognition of the role of counselor-assisted workouts should affect the debate over creditors' remedies, it is necessary to digress and first examine in some detail the rationale used to proscribe some of these remedies.

B. The Problematic Basis of the Lost-Value Premise as a Sole or Principal Justification for Abolishing or Restricting Certain Creditors' Remedies

(1) The Development of the Lost-Value Premise

Robert Scott has observed that the rapid growth in consumer credit in the post World War II years, characterized by fluid entry and exit of firms, often produced unstable conditions that were "a breeding ground for fraudulent and deceptive practices."[1] Scott notes, however, that "[a]s the consumer credit market stabilized, the incidence of fraud and deceptive practices declined."[2] He observes that the regulatory focus then shifted to the more subtle problem of lost value in the collection process. Simply put, the lost-value problem is that the injury to debtors caused by the use of various creditors' remedies "frequently seemed greater than the cor-

---

1. Robert E. Scott, *Rethinking the Regulation of Coercive Creditor Remedies*, 89 COLUM. L. REV. 730, 734 (1989).
2. *Id.*

responding benefits to creditors."³ The problem of lost value has emerged as the principal issue in the debate over reform of creditors' remedies.

In the materials that follow, I borrow extensively from the work of others, for those who have theorized on the concept of lost value and the effect that concept should have on provision for creditors' remedies have done significant work. My emphasis on competition among creditors as the overarching impetus to collection practices that may result in lost value builds on and often complements, I believe, the work of these earlier theorists.

While lost value was first attributed to creditors' failure to maximize their return by the use of spiteful coercive collection practices often coupled with deceptive practices in extending credit, Robert Scott credits the late Arthur Allen Leff with shifting "the debate from creditor misbehavior to the structural dynamics of the postdefault collection process,"⁴ which produces, according to some analysts, lost value.

In his important study, however, Leff did not address that aspect of lost value that stems from competition among creditors. He analyzed collections as contests between the debtor and her individual creditors, but not, additionally, as contests among those creditors.⁵ Using the terminology of game theory, Leff reasoned that when one takes account of the transaction costs imposed upon both the debtor and creditor of playing out the collection process, collection is not a two-person zero-sum game, but a two-person minus-sum game.⁶ The transaction costs imposed upon the creditor are the obvious volitional expenses incurred in the creditor's attempts to collect by self-help or judicial action.⁷ Coercive collection, however, may impose collection costs on a debtor who does not mount a defense to the creditor's judicial or self-help action.⁸ According to Leff, these costs arise when the passive debtor's property is taken and lost value results from that taking.⁹

Leff illustrates this concept of lost value by contrasting the value of a machine to a debtor who has knowledge of the machine's material peculiarities and how to adjust for them with the value of that machine to a potential purchaser at a creditor's disposition sale. The potential purchaser discounts the value of the machine because he knows that dis-

---

3. *Id.* at 734–35.
4. *Id.* at 735.
5. Arthur A. Leff, *Injury, Ignorance and Spite—The Dynamics of Coercive Collection*, 80 YALE L.J. 1, 5–26 (1970).
6. *Id.* at 6.
7. *Id.* at 7–10.
8. Id. at 10–18.
9. *Id.*

covering its peculiarities and learning how to adjust for them will require an investment of his human capital.[10]

Leff then lists additional factors that may lower the price of used goods in an aftermarket below that attributable to normal depreciation. His culprits are "restricted buyer pools approaching minimonopsonies, insufficient sale advertising, unenthusiastic and incompetent selling, title uncertainties, and the lack of price-maximizing incentives."[11] Moreover, Leff does not limit the application of the lost-value premise to forced sales of goods but also finds lost value in garnishment of wages. Here, the loss to the debtor over the gain to the creditor is found in the different marginal utility of money to those parties and the adverse effect of garnishment on employer-employee relations.[12]

Leff attributes the incidence of lost value and related transaction costs in the collection process to deficiencies in the flow of information between the debtor and creditor.[13] If information were perfect, creditors would not incur costs attempting collection from debtors who are "can't pays." Likewise, a debtor who would otherwise be a "won't pay" would find some means of settling his obligation when he learns that his creditor has cost-effective remedies. That settlement would enable the debtor to avoid any taking of his property that might result in lost value. Leff concludes that "[t]o the extent that actualized coercion is a source of waste in the coercive collection game, its replacement by successfully communicated information offers the prospect of greater efficiency and, depending on one's definition, perhaps more justice."[14]

Leff's prescription for curing the problem of lost value was to "have the government supply, in addition to the umpired killing ground of the current judicial coercive system, an impartial source of particularized reality and a conversation pit."[15] The government agency that Leff proposed would exert every effort to ensure that the parties actually confronted each other to discuss their dispute before coercive practices could be had.[16] Leff would inject an "impartial 'referee'"[17] into the process to "make known to the parties what the legal situation is."[18] Leff would use the resources of the state not just to determine the issue of liability, a

---

10. *See id.* at 12.
11. *Id.* at 13.
12. *Id.* at 15.
13. *Id.* at 26–36.
14. *Id.* at 26.
15. *Id.* at 43.
16. Id. at 43–44.
17. *Id.* at 44.
18. *Id.*

nonissue in most collection cases, but also to reduce the transaction costs of collecting claims or determining that they are non-collectible. The debtor would learn the extent of the creditor's arsenal of remedies, and the creditor would learn the limitations on his powers to collect based on the debtor's circumstances.

Leff justified the imposition of the costs that his procedure would entail by noting that those costs might be more than compensated for by the avoidance of other costs and losses.[19] He asserted that implementation of his proposal "would transform the payment procedure back to the voluntary mode from the coercive mode."[20] If an acceptable and feasible payment plan resulted from the process, the creditor's risks of not collecting would diminish for, among other reasons, the plan would avoid the problem of lost value.[21]

Leff viewed asymmetries of information among the debtor and creditor as the principal reason for the failure to substitute cooperative collection for value-destroying coercive collection. He observed that "[m]ost important, paying for accurate particularized information would lessen the chance of an unnecessary coercion game being played solely because of the parties' ignorance of the realities of the situation."[22] In a final statement addressing the potential merits of his proposed system, Leff momentarily expanded his horizons on the collection setting from one concerned with costs and benefits to the debtor and one creditor to one concerned with the subject of this work, costs and benefits to the debtor and all his creditors.

> It is even possible that the gross costs of collection would decrease, that the costs and losses not only to the C-D [one creditor and the debtor] "team" but even to Cs [creditors] as a group... are greater under the current system than they would be if more precise information were "bought."[23]

The similarities between the functions of the agency that Leff proposed and the functions of a consumer credit counseling agency are readily apparent. In their most important function, both a counseling agency and Leff's "refereed conversation pit" are designed to provide parties with reliable information upon which they may fashion a more effective and efficient and less painful method of debt payment. But unlike credit counseling, which addresses all claims against a debtor, Leff's conversation pit focused on the claim of a single creditor, and he did not describe how

---

19. *Id.* at 45.
20. *Id.*
21. *Id.*
22. *Id.*
23. *Id.*

his plan would deal with a debtor in arrears to multiple creditors. There are other notable differences in the conversation pit and a consumer credit counseling agency.

Leff's referee would not be subject to the limitations that give creditors reason to trust, without cost-prohibitive monitoring, the work of credit counselors. Unlike the credit counselor who is limited to proposing feasible extension plans for full payment of claims, Leff's referee would apparently have the power to reduce the creditor's claim based not only on legally recognized defenses such as breach of warranty but also on the debtor's ability to pay. In this respect, the referee would have greater powers than those of small-claims courts in some states that may stay execution while a financially strapped debtor makes installment payments on the full amount of the creditor's judgment.[24] These broad powers that Leff assigned his referees would doubtlessly concern creditors as they did Leff himself:

> Carrying out the referee's job with fairness and competence may be beyond the powers of mankind. The critical danger is that the referee would either fall into the trap of trying to settle everything, no matter how outrageous the creditor or debtor conduct, or decide that absolute "justice" demanded the exercise of all possible "rights" to the full. These difficulties are not to be lightly dismissed. After all, this is hardly the first time that a mediation role has been suggested as an innovative grafting onto a conflict model, and the results in other contexts are, to say the least, ambiguous.[25]

There is another feature of Leff's proposed cure for harmful coercive collection that differs materially from the remedy provided by counseling agencies. The debtor and creditors voluntarily submit to the work of a counseling agency, for these agencies, which operate in the private sector, have no other jurisdiction over collection cases. Leff proposed a public agency that creditors were required to use before they could resort to coercive measures. His choice of a public agency to remedy the problems of overly coercive collection is not free from ambivalence. In a footnote to his statement that the government should provide the remedial agency, Leff states:

> I am not sure why it must be the government. If trade fairs pay, why don't settlement fairs? It may only be that the legal framework of a capitalist economy, with its necessary public defense of private claims, already provides with its judicial-coercive model a collection system at public expense which *seems* so efficient that no private competitor can

---

24. *See e.g.* FLA. SMALL CLAIMS R. 7.210.
25. Leff, *supra* note 5 at 44–45.

survive. That is, the state is already intervening to make a particular kind of "market." Or it may be because the most lucrative business, that involving intra-business disputes, already has its private mechanisms. On the other hand, my leap to the government may just be a spasmic liberal gesture.[26]

To the extent that creditors and debtors view workouts through counseling agencies more favorably than bankruptcy because of the voluntary nature of the workout in a private-sector setting, this benefit would be lost under Leff's proposal.

Note that Leff failed to mention consumer credit counseling agencies in his work. This may be partly because he was concerned with disputed as well as undisputed claims, and counseling agencies play no significant role in resolving the former. But Leff may simply have dismissed the work of counseling agencies as insignificant. If so, his failure to recognize the role of these agencies was far more justifiable at the time he wrote than it would be today. In the latter 1960's, counseling agencies were far fewer in number and served a much smaller sector of the consumer credit market than they do today.

Criticism of Leff's proposed remedy for curbing coercive collection should not detract, however, from the significance of his work. He recognized that the dynamics of the collection process could produce lost value whether or not there was creditor misbehavior. His attribution of destructive coercive collection practices to a creditor's inadequate or mistaken information about the debtor also suggested a role for additional deficits in creditor information to play—the one that I have examined in this work. When a debtor in dire financial straits needs team effort by her creditors to effect a workout, each creditor needs to know that other creditors are committed and will stay committed to an extension plan that the debtor is performing.

It is worth noting that Leff's referred conversation pit could beneficially coexist with a system making extensive use of consumer credit counseling agencies. His referred services addressed resolution of disputed claims, something that counseling services do not. There are other points of complementarity. When the debtor, for whatever reason, fails to resort to a counselor-assisted workout or bankruptcy, Leff's referred conversation pit might be preferable, because of its ability to eliminate lost value, to proceeding immediately to judicial collection or credible threats to do so. In the case of a debtor who is in arrears to more than one creditor but whose financial plight is not hopeless, Leff's referee might be instrumental in convincing the debtor to engage the services of a credit counseling agency.

---

26. *Id.* at 43, n.134 (emphasis in original).

Leff's lost-value premise has significantly influenced the debate over creditors' remedies. William Whitford further developed Leff's ideas nearly a decade later, toward the end of the 1970's, in an extensive analysis of the rules and practices of consumer-debt collection. Whitford observed that, due to the costs to both debtor and creditor of coercive collection, consensual payments reached by post-default bargaining characterize and should characterize the collection of consumer debt.[27]

But Whitford found that asymmetries of information, competition among creditors and creditors' abilities to place costs of coercive collection on debtors or parties external to the transaction, such as taxpayers who provide safety nets for impoverished debtors, often result in coercive collection replacing the value-maximizing bargaining model.[28] Whitford maintained that the principal function of the rights given creditors and debtors under any legal collection procedure lies not in the direct use of those rights, such as creditors' employment of execution process or debtors' invocation of bankruptcy relief, but in the bargaining leverages the rights additionally provide.[29]

Unlike Leff, Whitford recognized the significance of competition among creditors in creating lost value. He observed that creditor competition may predispose an unsecured creditor to reject a refinancing agreement in favor of seeking quick execution upon property.[30] Like Leff, however, Whitford failed to note the presence of consumer-credit counselors as agents for taming unbridled, harmful competition among creditors.

In spite of the prominence assigned the lost-value premise by analysts Leff and Whitford, that premise is not without its forceful critics. Alan Schwartz has searched diligently for lost value in a setting in which it is commonly assumed to thrive: the self-help repossession and sale of consumer goods by secured creditors.[31] He finds that "if the harm that repossession imposes on debtors is conceptualized in economic terms... repossession imposes trivial harms, no harms at all, or harms that cannot be shown to exceed the gains, depending on how one considers the harm to have been inflicted."[32]

Schwartz examines and rejects four possible sources of lost value in the repossession and sale of consumer goods. First, the debtor may value

---

27. *See* William C. Whitford, *A Critique of the Consumer Credit Collection System*, 1979 WIS L. REV. 1047, 1048–72.
28. *See id.* at 1106–09.
29. *See id.* at 1048–49, 1051–1056.
30. *See id.* at 1066–71, 1077–78.
31. Alan Schwartz, *The Enforceability of Security Interests in Consumer Goods*, 26 J.L. & ECON. 117, 139–48 (1983).
32. *Id.* at 147.

the goods more than their market price. But this will lead to routine default and loss of the goods by a debtor only when a change in the debtor's circumstances, such as loss of income or the incurring of unanticipated expenses, causes the debtor to derive greater utility from retaining the unpaid portion of the purchase price than from retaining the goods.[33] Second, ineffective resale markets may depress value, but loss upon resale is often not attributable to repossession but to other factors, such as the premium commanded by new goods over used ones, whether or not the used goods are repossessed.[34] Third, human capital may be destroyed when goods must be operated in a certain fashion, but one not known to potential purchasers at a foreclosure sale. Schwartz concludes that these human-capital losses will be occasional and slight, however, for instances in which the debtor has invested substantial time in acquiring special knowledge relative to use of the goods that will require a like investment by the new owner are rare.[35] Fourth, lost value may result from the debtor's psychic losses such as the humiliation the debtor may suffer when creditors foreclose on collateral. But Schwartz is unconvinced by this argument as well, for, among other reasons, assuming the law must provide some means of enforcing contractual obligations, it is problematic that debtors would find an alternative collection remedy such as garnishment of wages to be less humiliating than repossession.[36] Even if one rejects Schwartz's measure of the triviality or absence of lost value in the quintessential paradigm for the phenomenon he examines, certainly his arguments should give pause to those who advance the conventional theory of lost value as a sole or principal reason for proscribing various creditors' remedies.

(2) A Different Perspective: The Benign Function of the Potential for Lost Value

Robert Scott also finds lawmakers' reliance on lost value as a broad justification for proscribing creditors' remedies to be misplaced, but his analysis proceeds on different grounds that do not deny that debtors' losses at the time their property is taken may in fact exceed creditors' gains. To the contrary, his justification for these remedies is predicated on their potential to produce lost value.

---

33. *See id.* at 140–44.
34. *See id.* at 144–46.
35. *See id.* at 147.
36. *See id.* at 147–48.

According to Scott, the potential for lost value produces a benefit by providing a means of reciprocal commitments that compensate for deficiencies in legal enforcement.[37] The lost value that the debtor may suffer from the creditor's use of the remedy to take property commits the debtor to pay the indebtedness. On the other hand, the market value of the property to the creditor, which is low compared to the value of the property as leverage against the debtor, removes any incentive of the creditor to resort to his remedy by inducing or fabricating a default after partial payment.[38] The low resale value of the property also serves to deter the creditor from trigger-happy seizure prior to giving the debtor a reasonable opportunity to cure any actual default.

The benign purpose of the potential for lost value then is to make mutually advantageous credit transactions, which might not otherwise transpire, doable. When a creditor believes that the interest rate offered a loan applicant gives the creditor a higher return than she can obtain elsewhere, and the applicant believes that payment of that creditor's proposed rate of interest is preferable to not borrowing or paying the rates offered by other creditors, the transaction may nevertheless fail to materialize if either party fears that the other may not perform.

Scott analogizes collateral with lost-value potential to the case of the medieval king who offers his sickly son, the "puny prince" as hostage to ensure the king's performance of an undertaking to a neighboring monarch. The king's love and affection for his son assure the value of the king's commitment, while the risk that the creditor monarch will provoke a default and make off with the hostage is greatly reduced by the prince's negligible value as a warrior.[39] Scott concludes that "[t]he puny prince theme suggests that for cases in which [the] [d]ebtor is offering assets as security, the ideal choices are family heirlooms, household goods, future human capital, or indeed any other entitlements that have substantial idiosyncratic or sentimental value to [the] [d]ebtor."[40]

Instead of focusing on the post default effects of creditor leverage, which to some appear suspect, Scott focuses on the potential for lost value at the time the parties are contemplating entering into the credit contract. At that time, that potential signals to each potential contractor the mo-

---

37. Scott, *supra* note 1, at 734–56.
38. *See id.* at 746–48.
39. *See id.* at 748–49. Scott attributes the original hostage imagery to Thomas Schelling and his pioneering work in applied game theory. *See id.* at 741, n.44; 748, n. 64 citing THOMAS SCHELLING, THE STRATEGY OF CONFLICT 135–36 (1960).
40. *Id.* at 749 (citation omitted).

tivation of the other potential contractor to perform the obligation the contract will impose on him.[41]

The puny prince metaphor aptly dramatizes the role played by lost value in consumer-credit transactions. The debtor as well as the creditor may need assurance of the other's performance, although the debtor's need for that assurance will arise far less often than the creditor's. Creditors, however, may induce or fabricate a default after partial payment because they perceive an economic advantage in doing so.

This may occur when creditors feel they can recover and retain more than the amount of their claims by repossession and sale of collateral, in spite of a long recognized rule of commercial law that allocates surpluses to debtors.[42] Here the lesson of the puny-prince metaphor for debtors who harbor concern about creditors' misbehavior is that the market value of the collateral should not significantly exceed the amount of the debt.

Another form of creditor misbehavior cannot be forestalled, however, by limiting the value of the collateral relative to the amount of the creditors' claim. This occurs when interest rates have risen so that the creditor can receive a higher return on the principal recovered after wrongful acceleration of a fixed-interest debt. But this temptation will only exert significant pull on creditors when interest rates increase substantially and creditors are financing large, long-term debts.

Moreover, courts are not disposed to look kindly on creditors who fabricate or induce defaults for purposes of repossession, premature recovery of principal or other reasons. Lender liability cases may result in large recoveries for debtors that exceed their compensatory damages. Punitive damage awards are common in wrongful repossession cases, and courts in those cases will let substantial jury awards stand, even when the creditor's conduct was not willful.[43] Thus, the possibility that the creditor may lose far more from his misbehavior than he may gain motivates him to faithfully perform the contract.

Creditors also have their reputation with customers and potential customers to consider. It is simply not in the economic interests of legitimate lenders to induce or provoke a default. Although some evidence exists of

---

41. *See id.* at 740.
42. *See* U.C.C. § 9–504(2) (1994).
43. *See e.g.* Mitchell v. Ford Motor Credit Co., 38 U.C.C. Rep. Serv. 1812 (Okla. 1984) (In a case in which debtors sustained only $843.74 in actual damages for wrongful repossession, the court rejected the argument that the $60,000 punitive damage award was out of proportion to the injury inflicted and held that the jury could conclude from the evidence that the flawed work process of the finance company demonstrated a reckless disregard for the rights of the debtors.).

creditors making and retaining profits on resale of collateral,[44] law and market forces generally control such practices among all but the most disreputable of creditors.

Because performance of the creditor's principal duty—the making of the loan—must necessarily precede performance of the principal duty of the debtor—repayment of the loan—the most frequently encountered impediment to mutually advantageous credit contracts is the creditor's fear that the debtor may not pay. When the loan applicant is a marginal credit risk, her ability to credibly commit herself to repayment is an especially valuable right. By agreeing to terms that may impose the penalty of lost value on her if she defaults, the applicant may obtain a badly needed loan that otherwise would not have been made. Thus, the debtor's grant of a security interest in property worth more to the debtor than to the creditor in the event of default is understandable in light of the discipline that both debtor and creditor justifiably believe it imposes on the debtor to perform her duties under the contract.

The penalty imposed upon the debtor by lost value diminishes the possibility that she will engage in conduct after credit is extended that endangers her ability to repay. Scott provides examples of such debtor misconduct. The debtor may invest the loan proceeds in high risk investments, gambling with the creditor's funds because any gains that materialize will be the debtor's, while losses will probably fall on the creditor. More likely in consumer lending, the creditor is concerned that after the loan is made the debtor may secure additional credit that will dilute the creditor's claim to the future earning potential and nonexempt assets of the debtor. If debt burden becomes excessive, it may also reduce the debtor's incentive to pursue income-generating activities because her creditors will derive much if not all of the benefits of her future efforts.[45]

Recognizing the need to make the debtor's promise of repayment credible, the common law with legislative assistance over the years has long provided a remedy for a creditor who starts the collection process with nothing more than a claim against her debtor. If this claim is one for which the law will provide relief, at least in theory if not in fact, a mere unsecured creditor, with no access to special remedies provided by the debtor-creditor contract, may eventually hope to obtain a judgment. The judicial proceeding that produces that judgment may entail, however, no small investment in time, attorney's fees and court costs relative to the

---

44. *See e.g.* Shuchman, *Profit on Default: An Archival Study of Automobile Repossession and Resale,* 22 Stan. L. Rev. 20 (1969); White, *Consumer Repossessions and Deficiencies: New Perspectives from New Data,* 23 B.C.L. Rev. 385 (1982).

45. *See* Scott, *supra* note 1 at 744–45.

size of the claim in many instances of action on consumer debt. Of course, before that judgment is rendered, the debtor is entitled to all the protection provided by our federal and state constitutions, statutes and rules of procedure—law that certainly in its general application properly favors accuracy and restraint over efficiency and expeditiousness. While cases in which a consumer debtor who is justly obligated resorts to dilatory actions or spurious defenses are uncommon, and the creditor in most cases can expect to take a default judgment, one rendered when the debtor fails to interpose a defense, going to court is not costless in either time or money.

Once obtained, a monetary judgment is not self-enforcing. Before the force of the state may be brought to bear to assist the creditor in collecting her claim, a writ of execution or garnishment authorizing seizure of the debtor's nonexempt property must be issued, and, here again, the law may provide for further delays. A debtor who has contested the creditor's claim may have additional time to file a motion for a new trial or for a rehearing, and he may appeal a judgment taken against him.

Although the writ of execution once issued commands the sheriff or other law-enforcement officer to seize nonexempt property of the debtor sufficient to satisfy the creditor's judgment, it is the creditor who must ferret out property upon which levy can be made. Further judicial action, proceedings supplementary to execution or discovery depositions, may be used to examine the debtor and others regarding the debtor's assets if nonexempt property cannot be located otherwise. But again, this action entails additional costs and delays. Locating property by nonjudicial means does as well, and regardless of the methods employed, the judgment creditor may fail to find any nonexempt property that may be subjected to payment of her claim. The uncertainty of locating nonexempt assets to satisfy a judgment obtained by resort to the basic judicial collection process is usually the most significant impediment to the use of that process in consumer collection cases.

Obviously, even before initiating suit, the prudent creditor will have assessed the probability of collecting a judgment. But the delay attributable to the unsecured creditor's inability to seize property of the debtor until after a postjudgment writ is issued may result in such changes in the facts upon which the earlier assessment was made as to defeat the creditor's favorable projections. In exceptional circumstances, such as when the unsecured creditor can prove that the debtor is fraudulently transferring, secreting, or removing his property from the court's jurisdiction, prejudgment attachment or garnishment statutes will provide for a provisional seizure of the debtor's nonexempt property prior to judgment. Instances in which the creditor has sufficient evidence of such conduct by a consumer debtor in time to prevent the loss of assets upon which the creditor's claim may be satisfied do not occur often, however, and prejudgment process imposes additional costs and risks on the creditor. She is required to post surety bond to compensate the debtor if either her grounds for the prejudgment writ or the basis for her underlying claim fail to pass subsequent judicial muster.[46]

If the judgment creditor is successful in having the sheriff levy upon nonexempt property of the debtor with more than nominal market value, she must incur still additional costs before receiving any return on her investment in collection effort. In addition to the costs of the levy, the creditor must advance funds for storage and preservation of the property levied upon, for the costs of notices of sale, and for the sheriff's fees in conducting the execution sale. These additional proceedings consume time as well as money, and although the levying creditor is entitled to recover her execution costs out of the proceeds of the execution sale, this recovery will reduce the net proceeds of that sale, which provide the only means of satisfying her judgment.

The foregoing examination of the principal features of the basic judicial collection process that the state provides for the enforcement of all claims reveals that it is fraught with uncertainty, costs and delays for creditors. This is not to say that providers of consumer credit do not go to court using nothing more than basic judicial collection procedures, for they do so in numerous cases filed every business day. But as noted earlier in examining creditors' strategies, creditors' investments in many of these suits may be justified in large part by the enhancement of their reputation with other debtors as collectors who do not quit easily. Failing efforts by direct contact to talk the debtor into paying, all too often the creditor's only recourse is to write off the debt as uncollectible.

Scott finds two factors in consumer-credit transactions that impair the efficacy of the enforcement of contracts. First, he notes the significant restrictions that state collection law places upon the creditor's ability to collect out of the debtor's assets. These restrictions result from exemption laws, which limit unsecured creditors rights to enforce their claims against certain property of their debtors, and limitations on prejudgment garnishment and attachment, which delay creditors' rights to seize nonexempt property and thereby provide debtors with a means to use strategic delay. Second, Scott finds that even these restricted collection rights under state law may be trumped by the debtor's decision to seek discharge in bankruptcy.[47]

Because resort to the procedures of the basic judicial collection process provided all creditors have often proved inadequate, creditors have long

---

46. *See* Winton E. Williams, *Creditors' Prejudgment Remedies: Expanding Strictures on Traditional Rights*, 25 FLA. L. REV. 60 (1972).

47. *See* Scott, *supra* note 1 at 741–44.

provided for additional remedies in their contracts with debtors. Particularly in consumer lending, where transaction-cost barriers cast long shadows over relatively small claims, creditors must seek ways to reduce the uncertainty, costs and delays of collection.

To avoid both exemption claims and delays in rights to seize and sell debtors' property, creditors have found no better device than security interests in their debtors' property. As noted previously in examining collection practices, exemption laws generally do not apply to enforcement of security interests, and secured creditors are commonly given the right to use prejudgment judicial process and even self-help to promptly seize their collateral upon default by the debtor. A security interest also reduces the uncertainty of collection by providing the creditor with advance knowledge of the existence of property to which she may turn in the event of the debtor's default. Earlier examination of contests among creditors revealed another significant benefit of a creditor obtaining a security interest. If the secured creditor's interest is perfected by filing or otherwise, her priority in the collateral is established against most competing interests that arise thereafter. Even in the debtor's bankruptcy, the holder of a perfected security interest is protected by the value of her collateral.

Ideally then, the risks, costs and delays of collection are best contained when creditors may resort to easily recoverable and readily resalable collateral with a market value sufficient to cover their claims and costs of repossession and resale. Large commercial creditors—banks that are principal lenders to their business debtors—are often particularly apt, due to the commanding position they occupy, at adequately collateralizing their loans. Providers of consumer credit, however, have far more limited sources of collateral to tap. The consumer's automobile, boat, furniture and appliances—so-called consumer durables—have value to the secured lender but often depreciate faster and are normally more readily removable from the creditor's grasp following default than the business firm's equipment. The commercial financer has access to collateral that does not depreciate from use or sale to an end-user, but in consumer lending there is no equivalent of a business debtor's inventory. While the consumer's home may retain its value as well as the firm's factory, store or office, the security interest creditors once took in consumers' wages did not diversify their risk as well as commercial lenders' interests in merchants' receivables. A security interest in wages could become worthless for the same reason that the consumer defaulted—loss of his job.

For their often unsecured or undersecured loans, many consumer lenders turned to other contractually created remedies that partook of some of the advantages afforded them by security interests. Contract terms containing waivers of exemptions once gave creditors in jurisdictions where they were permitted the right to seize property that otherwise could not be subjected to their claims, but these waiver of exemption terms provided no innate rights to prejudgment seizure as security interests do. Another supplemental, contractual remedy formerly used in some states, however, the cognovit or confession of judgment, gave creditors a means of expediting the process of obtaining a judgment, but no escape from the barrier of exemption laws. By cognovit or confession of judgment clause the debtor consented to the creditor obtaining a judgment without the debtor receiving notice and an opportunity to contest the creditor's claim in a court hearing until after the judgment was rendered.

While supplemental remedies may benefit consumers when they apply for credit by lowering its costs and increasing its availability, the factor that produces that benefit—consumers' increased commitment to obligation based on the penalty that may stem from default—is apt to make consumers who do default highly vocal advocates for legal reform. Some consumerists who took up the cause of these defaulting consumers were prone to focus primarily, if not exclusively, on debtors' post default losses and to denigrate, if indeed they discerned, the *ex ante* benefits of supplemental remedies. And with that focus, the use of security interests in some instances and all uses of waivers of exemption and confessions of judgment lent themselves readily to unfavorable scrutiny under the lost-value premise.

(3) Lawmakers' Attacks on Creditors' Remedies Based on the Lost-Value Premise

While commentators have differed on whether the lost-value premise justifies abolishing certain creditors' remedies, many lawmakers have accepted that premise as a sole or principal justification for removing many supplemental remedies from the arsenal of creditors. Scott has summarized the role of lost value in legislative and administrative attacks on these remedies.

> Th[e] lost-value premise formed the basis for a widely accepted hypothesis that both the threat and the exercise of self-help remedies would, absent legal intervention, lead to normatively undesirable coercion and exploitation of consumer debtors…In response, a number of states enacted special legislation regulating self-enforcing remedies. Some states chose to ban deficiency judgments, [which represent the balance remaining after the net proceeds from sale of collateral are credited to the debtor's account,] in certain sales transactions, while others limited the enforcement of wage assignments, waivers of asset exemptions and confession of judgment clauses. Finally, in 1984 the Federal Trade Commission ("FTC") promulgated a Credit Practices Rule declaring most self-enforcing remedies "unfair" to

consumers. The FTC relied specifically on the lost-value premise in banning contract terms providing for blanket security interests in household goods, wage assignments, waivers of asset exemptions and confessions of judgment.[48]

Scott's demarcation of the creditors' remedies that have fallen victim to the lost-value premise is couched in terms of creditors' "self-help remedies" and "self-enforcing remedies." But he intends these terms to have broad meaning, for his enumeration of banned remedies includes those that result from terms in the loan agreement that either require or may require judicial action to enforce. Obviously, a term in a loan agreement providing for confession of judgment or one providing for waiver of exemptions was useful, aside from its value as leverage, only if legal action was taken. A secured creditor could formerly seize all the debtor's household goods under a blanket security interest without resort to judicial action if this could be done without breach of the peace. But if the debtor rebuffed the creditor's efforts at self-help repossession, a writ of replevin providing the assistance of law enforcement officers would be needed to take possession of the collateral. States that banned deficiency judgments in certain sales transactions proscribed the only step in the collection of a claim by a secured creditor that requires judicial action in all instances. Creditors did realize upon wage assignments prior to their ban by simply notifying the debtor's employer of the assignment, and no court action was required unless the employer failed to recognize the assignment. Most of the remedies with which Scott is concerned, however, are creditors' "self-help" or "self-enforcing remedies" only in the sense that they remove some of the obstacles that face a creditor using ordinary judicial collection process and are created by creditors' actions in contracting for the remedy or a security interest that includes the remedy. Scott sheds further light on the breadth of the attack on creditors' remedies that he chronicles by noting that these remedies are, "at least to some extent, prior to and independent of the alternative process of postjudgment execution."[49]

As the battles over creditors' remedies since the beginnings of the current wave of consumerism in the 1960's have raged over contractual terms that provide for remedies that supplement those the law provides for the

---

48. *Id.* at 735–36 (citations omitted). Scott's statement that the FTC relied specifically on the lost-value premise in promulgating the Credit Practices Rule is supported by no less than three citations in his work to the FTC's Statement of Basis and Purpose and Regulatory Analysis. *See id.* at 731 n.5, 736 n.24, 737 n.28, citing FTC Trade Regulation Rule: Credit Practices, 49 Fed. Reg. 7740 at 7779–81, 7743–45 and 7762–65 (1984), respectively.

49. Scott, supra note 1, at 730–731, n.2.

enforcement of all monetary obligations, the obvious question is whether the costs of these remedies to debtors who default exceed their benefits to creditors and, by extension, to debtors who pay their obligations. In the Statement of Basis and Purpose and Regulatory Analysis of the Credit Practices Rule, the Federal Trade Commission (FTC) found that confessions of judgment, waivers of exemptions and the taking of security interests in two instances imposed greater costs than benefits on consumers.[50] Also of significance in the promulgation of the Credit Practices Rule, however, was the FTC's rejection of three provisions in the rule it had proposed nine years earlier that would have diminished the value of security interests in certain instances.[51] The discussion that follows examines first the remedies that the FTC banned and then those that it left to the discretion of the states.

The FTC found that although there were procedures for reopening confessions of judgment, the absence of notice and a hearing prior to entry of judgment caused significant consumer injury. Of concern to the FTC were the unknowing waiver of due process rights, the difficulties that might be encountered by a debtor who had a defense in successfully reopening a case, and the imposition of a judicial lien on the debtor's property without the debtor being aware that such a lien existed until she attempted to mortgage or sell the property. When these costs, especially those relating to loss of due process rights, are compared with the principal creditor benefit—avoidance of a some 30 day delay in taking a default judgment in an uncontested case—confessions of judgment are exposed as one of the more vulnerable of the supplemental remedies.[52]

To the extent that exemption laws provide debtors and their dependents with the necessities of life and no excessive amount of its frills at the expense of others, a case can be made against waivers of exemption on the basis of lost value. But the case is a far stronger one if the cost-benefit analysis is limited to the effects of the remedy at the time it is used or threatened to be used.

As a security interest in property will immunize a creditor against a claim of exemption in that property, the FTC states its case against waivers of exemption by considering the effects of such waivers on one category of exempt property that the FTC also insulated from the attachment of certain security interests—household goods. In its assessment of consumer injury from waivers of exemption, the FTC contrasts the economic, psychological and sentimental value to the debtor of household goods,

---

50. FTC, Trade Regulation Rule: Credit Practices, 49 Fed. Reg. 7740, 7744–45 (1984) [hereinafter FTC Rule].
51. *See id.* at 7783–84, 7786.
52. *See id.* at 7753–54.

which are a common exemption under state law, with the trifling resale value of these goods to the creditor. Because of the low resale value of household goods, the FTC notes that this property was rarely seized but, instead, provided a basis for *in terrorem* collection practices.[53]

But this is the very use that can be expected of a remedy designed more to ensure debtors' commitment to their obligations than to furnish a basis for recovery of the creditor's claim if default nevertheless occurs. As recovery of judgment and levy on household goods usually cost creditors more than they could recover from sale of those goods, creditors were prone to exercise their rights against household goods under waivers of exemption only when they thought debtors with the ability to pay were simply trying to avoid payment. In those instances, a creditor paid the costs of judgment and levy to maintain or enhance her reputation with other debtors as a creditor who would not easily concede enforcement of her claims. In cases in which debtors had plainly suffered some misfortune, such as illness or job loss, that significantly affected their present ability to pay, creditors did not need to incur judicial costs for the maintenance or enhancement of their reputations, and household goods were usually left with the debtor.

The FTC dismisses the argument that waivers of exemption are necessary to motivate some debtors to pay and the further argument that some exempt property has economic value to creditors with its finding that most debtor default is the result of factors beyond the debtor's control.[54] This finding fails, however, to address the many cases in which all or some significant part of the reasons for default are attributable to factors over which the debtor does exercise some meaningful measure of control.

Probably the most prevalent prior practice proscribed by the FTC rule was the taking of blanket security interests in household goods. Here the FTC prohibited the securing of credit with household goods when that credit did not arise from the acquisition of the goods in a credit sale or from the proceeds of a loan that was used to acquire the goods. These non-purchase money security interests usually covered all the debtor's furniture, appliances and other household goods as well as personal effects of the debtor and dependents of the debtor. The pawn shop was not targeted by the rule, as the proscribed security interests did not include those in which the creditor took possession of the collateral at the time the loan was made. Nor did the rule prohibit the taking of non-possessory non-purchase money security interests in certain household goods that were likely to have market value, such as works of art and jewelry.[55]

---

53. *See id.* at 7769–70.
54. *See id.* at 7770.
55. *See* FTC Credit Practices Rule, 16 C.F.R. §§ 444.1(i), 444.2(a)(4) (1996).

The fault the FTC found with these blanket security interests in household goods was essentially the same fault it had found with waivers of exemptions, when those exemptions include, as they frequently do, household goods. In fact, the blanket security interest in household goods was a more effective remedy than the waiver of exemptions, for it gave creditors the right to seize the goods prior to judgment and without judicial process if a self-help repossession could be accomplished without breach of the peace.

Again, the FTC stressed the difference in value of household goods in the resale market and their replacement, sentimental and emotional value to debtors. The record reflected that creditors rarely ever engaged in actual repossession of household goods, and that the occasional act of seizure appeared to be undertaken for punitive or psychological deterrent effect.[56] The FTC cites a passage from a finance company's manual to illustrate the real value of this collateral to creditors.

> Chase and recheck is a psychological device in which the Dial office representative visits the uncooperative customer's home specifically for the purpose of rechecking the security...Normally this will arouse concern on the part of the customer as for the reason for the rechecking. You are not to threaten that your branch is ready to repossess the security, merely advise the customers that you do not know the reason for the recheck, that you are just carrying out an assignment, and that if you were in similar circumstances you would contact the office immediately.[57]

The FTC found that the pressure creditors were able to exert on debtors by threat of repossession of household goods led to repayment arrangements that debtors would otherwise not make and that might worsen their situation. In particular, the FTC noted that that pressure could compel debtors to agree to refinance overdue obligations. The FTC explained that while refinancing would reduce or defer a debtor's scheduled monthly payments, "it does so by increasing the overall amount a debtor owes."[58] In light of the difficulty a debtor consciously bent on workout often encounters in obtaining extensions from one creditor without a means of assuring that creditor that the debtor's other creditor's are making similar concessions, is the FTC seriously suggesting that paying interest for an extension is usually not in the debtor's best interest?

In rejecting the consumer finance industry's argument that prohibiting blanket security interests in household goods would result in increased

---

56. *See* FTC Rule, *supra* note 50 at 7763–65.
57. *Id.* at 7764.
58. *Id.* at 7765.

The Significance of the Resolution of the Creditor's Dilemma 161

delinquency and/or foreclosure of high risk consumers from the credit market, the FTC analyzed data that revealed no significant difference between secured and unsecured borrowers' income levels and no significant difference between the level of indebtedness of such borrowers at the time credit was extended. That data did reveal, however, a higher average loan amount for secured loans than unsecured ones.[59] In justifying its proscription of blanket security interests in household goods, the FTC relied again on its finding of the principal cause of consumer default. "Given that the majority of defaults occur for reasons beyond the borrower's control, a threat to seize furniture and personal possessions is of marginal value in cases of serious delinquency."[60] The FTC does not address, however, those cases in which the debtor does exercise some measure of control over his financial affairs and the threat of seizure occurs in time to motivate the debtor to alter his financial course and avoid more serious problems.

In a further proscription of creditors' remedies under the Credit Practices Rule, the FTC banned wage assignments unless the assignment is revocable at the will of the debtor, constitutes a payroll deduction or preauthorized payment plan, or is an assignment of wages already earned.[61] Although assignments of wages are excluded from the Uniform Commercial Code's provisions for secured transactions,[62] because the sponsors of the Code recognized that "[s]uch assignments present important social problems whose solution should be a matter of local regulation,"[63] an assignment of wages does not differ in other respects from other security interests in receivables—amounts owed the debtor by her debtors.

When the collateral is receivables, the secured creditor need not sell the collateral upon her debtor's default but may realize its value by collecting from those parties who are indebted to her debtor. Thus, the creditor with a wage assignment could avoid sale of collateral in the depressed resale markets common to realization upon security interests in consumer goods. Because every dollar of debtor loss in the collection of a receivable is a dollar of creditor gain, the usual calculation of lost value—difference in the value of the property to the debtor and its value to the creditor—resulted in no loss in the case of wage assignments. William Whitford has argued that the common collection device of repossession and sale

---

59. *Id.* at 7766.
60. *Id.*
61. FTC Credit Practices Rule, 16 C.F.R. § 444.2(a)(3) (1996).
62. U.C.C. § 9–104(d) (1994).
63. Official Comment 4 to U.C.C. § 9–104 (1994).

of tangible collateral is unfortunate, and that execution on cash resources, particularly wages, should be facilitated.[64]

There is, however, a difference in garnishment of wages following judicial action resulting in a judgment for the creditor and the creditor's ability in a wage assignment to subject wages to her claim by simply giving the debtor's employer notice of the assignment. And the FTC seizes upon that difference in making its case for consumer injury by observing that "[w]age assignment, unlike garnishment, occurs without the procedural safeguards of a hearing and an opportunity to assert defenses or counterclaims."[65] But prejudgment taking of tangible collateral by self help is common practice for secured creditors, and when the secured creditor uses judicial process, she is commonly given the right to effect a prejudgment seizure, provided the debtor may subsequently contest that seizure. Moreover, the Uniform Commercial Code gives the secured party whose collateral is receivables the right upon default to collect from his debtor's customers without obtaining prior judicial approval.[66] By objecting to the creditor's right to commence collecting from the debtor's employer without a prior judicial hearing in which the debtor could raise defenses, the FTC ignored basic differences in remedies of secured and unsecured creditors, which have been recognized by the Supreme Court in cases in which they were subjected to the test of procedural due process.[67]

One significant difference does exist, however, in secured creditors collecting corporate debtors' accounts and secured creditors collecting consumers' wages. Many consumers, especially those who assign their wages as collateral, live from paycheck to paycheck, and any interruption of that income is likely to have dire consequences for them. While the FTC found that some states that allowed wage assignments had enacted limitations on the amount of wages that could be assigned, it concluded that these provisions were inconsistent and did not always offer adequate protection.[68] Thus, the FTC found consumer injury in the disruption of family finances that made it difficult for debtors to purchase necessities.

Other significant injuries to consumers that were found to result from wage assignments included interference with employment relationships,

---

64. William C. Whitford, *The Appropriate Role of Security Interests in Consumer Transactions*, 7 CARDOZO L. REV. 959, 960 (1986).

65. FTC Rule, *supra* note 50, at 7757.

66. *See* U.C.C. § 9–502 (1994).

67. *Compare* Sniadach v. Family Finance Corp, 395 U.S. 337 (1969) (unsecured creditor using prejudgment garnishment) *with* Mitchell v. W.T. Grant Co., 416 U.S. 600 (1974) (secured creditor using prejudgment process).

68. FTC Rule, *supra* note 50, at 7756.

because employers were hostile to wage assignments, and harm to consumers who had a defense to their creditor's claim, such as breach of warranty or fraud. Although the laws of some states gave debtors a means to stop payment of their wages to their creditors by giving notice of a defense to their employers, the FTC found that many consumers did not know how to avail themselves of this protection.[69] In finding that the offsetting benefits did not equal the harm to consumers, the FTC rejected the argument that wage assignments were necessary for poor credit risks who had no assets other than their paychecks to offer as collateral. Reliance was placed on postjudgment garnishment serving as a reasonable substitute for wage assignment.[70]

In its analysis of the projected costs and benefits of the Credit Practices Rule as a whole, the FTC considered comprehensive econometric analyses of creditors' remedies, interest rates, and amount of credit extended, which had been prepared for the rulemaking record. While doubtlessly the FTC had hoped that these econometric analyses would provide a more precise projection of the effects of its rulemaking action on the cost and availability of credit than could be obtained from other sources, such was not the case:

> We have given careful consideration to the econometric evidence assembled on this record... Our review has led us to consider that the econometric evidence does not, of itself, permit a definitive finding concerning the net costs or benefits of the rule as a whole. The relatively small magnitude of effects indicated by econometric evidence does permit us to be reasonably certain that the effect of the rule will not be unduly large in either direction. Our conclusion in this regard has led us to look more closely at the other available evidence on the rule as a whole and as to each provision.[71]

Turning from the econometric evidence, the FTC looked to the large body of experience with restrictions on creditors' remedies provided by comparisons between credit market conditions in states with laws similar to its proposed rule, a more stringent measure than the one promulgated, and other states. Most states already had laws similar to one or more of the provisions of the rule as promulgated and three states were identified that had legal regimes comparable to the earlier, more restrictive proposed rule. The FTC found that "[a]lthough there is some state to state variation, these comparisons reveal no systematic differences be-

---

69. *See id.* at 7757–59.
70. *See id.* at 7759–60.
71. *Id.* at 7780.

tween states that restricted remedies and the other states."[72] These statistics on credit markets in different states were bolstered by comments and testimony by state regulators and creditors from states that had adopted laws similar to the rule. Although occasional negative effects were noted, the FTC concluded that "[n]o significant effect on the cost or availability of credit was reported."[73]

In advancing a further argument that the remedies it banned would have little effect on credit practices because factors other than those remedies predominately determine costs and availability of credit, the FTC may have partially undermined its reliance on comparison of credit markets in various states.

> The most important factors [in determining costs and availability of credit] are: (1) the cost of money to the creditor, (2) the consumer's present income, existing debt level and capacity to incur further debt, (3) the possibility of the consumer being a repeat customer, (4) the creditor's opportunity cost, (5) the applicable interest rate ceilings, (6) the availability of other fees and charges, (7) the availability of the most useful creditor remedies, (8) the principal amount of the loan, etc[74]

If changes in the myriad of other variables mask the effect of loss of certain creditors' remedies, these state-to-state comparisons may fail to significantly reflect costs attributable to loss of those remedies.

As a final justification for its finding that all remedies proscribed by the rule "are ones whose importance to creditors is limited",[75] the FTC specifies two types of evidence of the value of remedies that relate to all its rule provisions. These two—evidence of the causes of consumer default and survey evidence on the importance of various collection methods to creditors—provide a basis from which the FTC "drew general inferences concerning the costs to creditors (and thus ultimately to consumers) of banning remedies."[76]

The FTC's evidence on the causes of consumer default was drawn from two studies, one by sociologist David Caplovitz and the other by the National Commission on Consumer Finance (NCCF).[77] In the Caplovitz

---

72. *Id.*
73. *Id.* at 7780–81.
74. *Id.* at 7781.
75. *Id.* at 7781.
76. *Id.*
77. *Id.* at 7747 citing DAVID CAPLOVITZ, CONSUMERS IN TROUBLE: A STUDY OF DEBTORS IN DEFAULT 54 (1974) and NCCF TECHNICAL STUDIES, Vol. V at 5 (1972).

study, based on a survey of debtors, loss of income from adverse employment changes was given as the leading reason for default, but voluntary overextension and debtor irresponsibility were given as either major or contributing causes of default by approximately one out of three of the debtors surveyed.[78] This category of debtors who by their own admission must share at least some of the blame is probably larger as many of the debtors who failed to live up to their obligation after they became separated or divorced also acted irresponsibly, and martial instability, a separate category, was given as a cause of default by eight per cent of the debtors.[79] Moreover, the response bias of debtors would tend to underestimate their own responsibility in causing default, and the categorizations of causes based on the information supplied by debtors were necessarily imprecise.

In the NCCF survey, which was one of creditors, unemployment took first place as the major cause of default reported by banks and retailers, but only tied for that position with overextension (apparently of the voluntary type as involuntary overextensions were subsumed by the NCCF's other categories) in the responses of finance companies. Overextension was reported as the second major cause of default by banks and the third by retailers.[80]

In spite of this evidence indicating that a significant number of debtors did not default simply due to conditions beyond their control, the FTC in numerous instances in its analysis of the Credit Practices Rule concludes that "default is largely beyond the debtor's control,"[81] apparently equating findings of a majority of the causes of default with substantially all such causes. In their recent commentary, Harvard's Elizabeth Warren and Jay Westbrook of the University of Texas have concluded that there is a case for the proposition that credit irresponsibility, and not just conditions beyond debtors' control, is a major cause of bankruptcy. They relate that consumer debt during the 1980's doubled from $300 to $600 billion, and conclude that "[g]rasshopper consumers willing to live on the plastic edge of financial collapse have met a credit industry all too willing to oblige."[82] While the corporate downsizing of recent years has resulted in many consumers having less income to meet existing obligations, a 1996 survey by Visa, the credit card company, found that nearly 29 percent of indi-

---

78. *Id.* at 7747–48.
79. *See* CAPLOVITZ, *supra* note 77 at 53, 85.
80. *See* FTC Rule, *supra* note 50, at 7748, note 31.
81. *Id.* at 7781. *See also id.* at 7748, 7766, 7770.
82. ELIZABETH WARREN & JAY LAWRENCE WESTBROOK, THE LAW OF DEBTORS AND CREDITORS 403 (3rd ed. 1996).

viduals recently filing for bankruptcy blamed their own excessive spending rather than family or professional emergencies for their plight.[83]

The FTC's second source of evidence of the minimal cumulative value of the remedies it proscribed in the Credit Practices Rule is drawn from a survey of the relative importance of various methods of collection to creditors taken by the National Commission on Consumer Finance. Based on that survey, the FTC concludes that "[t]he most important remedies—garnishment, repossession, acceleration of the debt, suit and direct contacts with debtors—are not restricted by the rule."[84]

It is not surprising that creditors would prefer loss of the remedies banned by the Credit Practices Rule to loss of the fundamental right to redress infliction of injury or loss in a forum provided by the state—the right of suit. Nor is it noteworthy that creditors value the remedy of garnishment, which provides one of the two principal means—the other is execution on tangible property—of bringing the force of the state to bear in enforcing a monetary judgment resulting from that basic right to sue. Of what use is a security interest, a creditor's principal means of elevating herself above the basic collection process, if the creditor has no right to repossess the collateral upon the debtor's default? And if creditors were forbidden the use of acceleration clauses in installment payment contracts—clauses which simply advance the maturity date of the unpaid principal upon the debtor's default—would installment payment contracts exist? Without an acceleration clause in such a contract, creditors would have to either bring a separate suit on each installment as it came due or wait until all installments were due before bringing one suit. Finally, would anyone seriously advocate the abrogation of the right of parties to a contract to try to resolve their differences by conferring with one another?

Certainly one may be skeptical of any argument that demeans the value of supplemental remedies simply by extolling the value of more fundamental ones. Scott, for one, questions whether the impact of the Credit Practices Rule on consumer finance is minimal.

> The FTC rule represents a significant expansion in the scale of regulation. While virtually every state previously had regulated some aspect of coercive collection, no state had prohibited the entire family of terms now banned by the FTC rule. Furthermore, the various state prohibitions did not seem to fall in any discernible pattern. In virtually every

---

83. Michelle Singletary, *A New Breed of Debtor Shocks Credit Card Issuers: As Delinquency Rates Hit Record Highs, It's Affluent, Low-Risk Consumers Who Are Declaring They Can't Pay*, WASH. POST, Sept. 18, 1996 at F01.

84. FTC Rule, *supra* note 50, at 7781.

The Significance of the Resolution of the Creditor's Dilemma 167

state, at least one of the now-prohibited contractual remedies was widely used, although the particular remedy of choice varied from state to state.[85]

Scott gleaned support for his appraisal of the effects of the Credit Practices Rule from the voluminous data the FTC supplied in promulgating the rule. For ease in reading the facts supporting Scott's appraisal, which for some inexplicable reason was relegated to a footnote, I have substituted footnotes of my own for citations in the text of his note[86] to the Statement of Basis and Purpose and Regulatory Analysis of the FTC's rule.

> There is substantial evidence that provisions for wage assignments, security interests in household goods, cognovit clauses [providing for confessions of judgment] and exemption waivers function as substitutes for one another. For example, in California, where repossession and resale were regulated, wage assignments were prevalent.[87] In New York, where confession of judgment was regulated, waivers of exemption were widely used (75% of the cash loan contracts),[88] as were wage assignments (68% of the small loan contracts).[89] Of these self-enforcing terms prohibited by the FTC rule, the most prevalent were security interests in household goods. Results of a survey of some 10,000 consumer accounts revealed that 76% of the precomputed loan contracts contained clauses authorizing household-goods security interests.[90]

The FTC's oft-stated concern with the impact of the rule on the costs and availability of credit certainly manifests its intent to look beyond lost value occurring at the time a remedy is used to the worth of that remedy as a deterrent of default as well as a device for recovery of claims when deterrence fails. The FTC has not proved, however, that on that basis costs outweigh benefits to consumers for the remedies it proscribed. My purpose is not to judge what I believe to be a judgment call by the FTC, one that may have involved factors that I have yet to explore. I do wish to add, however, another dimension to those that now frame regulatory decisions on creditors' remedies and will pursue that subject and explore additional concerns that drive the regulatory movement in the next section of this chapter. I am aware of only one statement by the FTC that addresses directly the subject of deterrence of default in the agency's analysis of the remedies it banned and, remarkably, that statement may be read

---

85. Scott, supra note 1, at 736 (citations omitted).
86. *Id.* at 736–37, n. 26.
87. FTC Rule, *supra* note 50 at 7757.
88. *Id.* at 7769 n. 10.
89. *Id.* at 7757.
90. *Id.* at 7762 n.12.

as tying deterrence value to the restraint the FTC exercised in failing to promulgate some of the bans on additional remedies contained in the rule the agency had proposed nine years earlier. "The remedies subject to the rule must be evaluated in light of their more limited incremental contribution to deterring default or reducing other creditor costs, given remedies that remain available."[91] Thus deterrence of default may also have played a role in the FTC's decision to preserve certain remedies. As the remedies the FTC left unscathed have been and will continue to be candidates for further regulation by the various states, the FTC's proposed but not promulgated bans merit consideration in any work concerned with creditors' remedies.

As previous analysis has shown, the hallmark of protected status for creditors is a security interest, for collateral may enable creditors to avoid both uncertainty and delay in collecting their claims. But a claim is a secured one only to the extent of the value of the collateral, and where that value is insufficient to cover the claim, the creditor is relegated for the deficiency to the status of an unsecured creditor. To recover that deficiency, the creditor must obtain judgment, seek other property of the debtor and subject it to seizure and sale through ordinary post-judgment judicial process. Secured creditors assume the risk that proceeds from the sale of collateral will be deficient to cover their claims, but they ordinarily do not assume the risk that they may not seek recovery of this deficiency through judicial process.[92]

The FTC considered but rejected a rule that would have required creditors with security interests in household goods, which were not prohibited because they secured purchase-money credit, to choose between repossession or suit on their claims. The rejected election-of-remedies rule would have forced the creditor to either take collateral worth less than her claim in full satisfaction of that claim or forego her interest in the collateral entirely and proceed against the debtor as an unsecured creditor. The FTC observed that this election-of-remedies rule would provide benefits to consumers similar in nature if not magnitude to those that result from the prohibition of non-purchase money security interests in household goods, but then concluded that the reduction in the value of collateral that would result from that rule would produce costs in excess of benefits.[93] The FTC offered no explanation of why cost-benefit analysis produced different results in the two cases, but purchase money security interests in appliances and furniture recently sold to the consumer are more likely to have resale value to the secured creditor than most prop-

---

91. *Id.* at 7744–45.
92. U.C.C. § 9–504(2) (1994).
93. FTC Rule, *supra* note 50 at 7784.

erty covered by blanket security interests in all the debtor's household goods.

Another proposed but rejected provision that would have had significant impact on secured creditors was a requirement that collateral other than household goods be valued at its retail price for purposes of calculating deficiencies. The FTC found that although sizeable deficiencies occur in the majority of transactions involving automobile repossessions, creditors pursue a deficiency only infrequently and recover no more than 5 to 15 percent of the deficiency. Because creditors are motivated to obtain the best possible price for collateral, the FTC concluded that there was insufficient evidence that valuation problems were prevalent.[94] Moreover, the FTC took notice of an elementary marketing principle: the difference between retail and wholesale prices is the result of the costs of retailing. Addressing another facet of this issue of valuation, the FTC found that while not all sales below wholesale book value could be explained by differences in the condition of the repossessed car and the average vehicle of its make and model, undervaluations were already addressed by the Uniform Commercial Code's requirement of commercially reasonable disposition, were not prevalent and could be dealt with on a case-by-case basis.[95]

A third restriction on secured lending in the proposed rule that was not promulgated by the FTC addressed the use of cross-collateral terms in credit-purchase contracts. Those terms provide that property purchased from a merchant secures not only the credit extended for its purchase but also certain credit extended the customer for prior and subsequent purchases. Under this arrangement, a customer's default in paying the balance due on the last item purchased would trigger the merchant's right to repossess all property sold the customer since her account had last been paid in full. The proposed rule would have reduced the amount of the retailer's collateral by requiring first-in, first-out accounting for credit contracts covering multiple purchases and thereby releasing collateral as the credit extended for each item purchased was paid in the order in which those items were purchased. The FTC found insufficient evidence of consumer injury from cross collateralization but also noted that the proposed rule could significantly reduce the value of purchase money security interests.[96] This issue has recently resurfaced, however. Section §9-104(d) of the August 7, 1997 Reporters' Interim Draft of the Revision of Uniform Commercial Code Article 9 provides that "payment must be applied to obligations secured

---

94. *Id.* at 7783.
95. *Id.* at 7784.
96. *See id.* at 7786.

by purchase money security interests [in consumer goods] in the order in which those obligations were incurred." While this provision represents the opinion of the reporters only, if the drafting committee and then sponsoring organizations of the Uniform Commercial Code adopt it, state legislatures following their recommendations will make inroads on the utility of cross-collateral terms in consumer finance that the FTC rejected.

Is it possible to reconcile the inroads that the FTC made on the use of secured credit with those that it considered but rejected? Two of the rejected proposals—election of remedies in the case of purchase money security interests in household goods and restrictions on the use of cross collateralization—relate only to purchase money collateral. The third restriction on rights resulting from security interests that the FTC rejected—valuation of collateral other than household goods at retail prices for purposes of calculating deficiencies—was directed primarily at the sale of new and used motor vehicles and therefore would have had its principal impact on purchase money credit as well. A purchase money secured creditor selling collateral in which he or an associate likely deal and that the debtor likely acquired only within the last year or so before repossession is likely to be selling goods with more than nominal market value. But the repossessing non-purchase money secured creditor with the now proscribed blanket security interest in household goods, which are defined in the Credit Practices Rule in such manner as to exclude most property of market value, was likely to have garnered little of interest to anyone in the resale market other than the most avid garage-sale shopper. The distinction between secured creditors' remedies that were banned by the FTC and those that were not would appear to be based then on the application of a relaxed lost value premise, but nevertheless one that was applied not at the time of application for credit but at the time the collateral was repossessed and sold. The FTC tolerated some lost value—the debtor might value the collateral more than the creditor could obtain for it on resale—so long as the creditor might derive some real market worth, and not just leverage, from the security interest.

Of course this theory fails to explain the FTC's proscription of security interest in wages, which are the only collateral in which it can not be argued, if we disregard different rates of marginal utility, that the debtor's losses exceed the creditor's gains. Here, as reported earlier, the FTC relied on a Supreme Court decision that addressed procedural due process rights in a prejudgment garnishment of wages by an unsecured creditor. The FTC failed to recognize the greater rights given secured creditors to seize collateral without prior judicial action. Nevertheless, debtors' dependency on uninterrupted flow of their paychecks do make wages, in the words of the Supreme Court, "a specialized type of property presenting distinct problems in our economic system."[97] But protection from acute loss of wages by wage assignments could have been provided in the same manner that some wages are sheltered from post-judgment garnishment. Most states exempt some part of the debtor's wages, and Congress has placed a ceiling on wages subject to garnishment for ordinary marketplace debt of 25 per cent of the debtor's earnings after all deductions for withholding required by law.[98]

In this section I have examined the actions of the FTC because that agency's prohibition of remedies has had effect nationwide, but historically, creditors' remedies have been defined by the states without assistance from Washington. With the exception of the establishment of a federal minimum exemption for wages and certain regulation of nonjudicial collection practices, the states had been subject only to 14th Amendment procedural due process oversight by the courts prior to the promulgation of the Credit Practices Rule in 1984. The FTC's action was significant then not only for its substance but also because it marked the entry of federalization into an area previously left solely to the states. The definition of creditors' remedies will no doubt remain, however, principally the domain of the states, and any work concerned with those remedies must necessarily concern itself with state law.

Obviously any attempt to examine in detail the laws of each of the states is beyond any reasonable scope for this work and certainly beyond the patience of its readers, but a compilation of the principal restrictions on remedies that have commanded the attention of those who wish to influence state legislators is provided by an act sponsored by the National Conference of Commissioners on Uniform State Laws for adoption in the various states—the Uniform Consumer Credit Code (UCCC). Unlike another act sponsored by those commissioners that relates to the law of transactions in the marketplace and that has been universally adopted by the states—the Uniform Commercial Code—the UCCC, in both its earlier (1968) and later (1974) versions, has been adopted by only 11 states.[99] But that is nevertheless a significant minority and, other states have adopted some of the UCCC's provisions in direct or modified form. Probably of greater import, the UCCC continues to provide a model for those interested in restricting creditors' remedies in whole or in part in the remaining states.

The UCCC restricts the taking of security interests in important instances in which the FTC's Credit Practices Rule does not. In contrast to

---

97. Sniadach v. Family Finance Corp. 395 U.S. 337, 340 (1969).
98. *See* 15 U.S.C. § 1673 (1994).
99. *See* 7 U.L.A. 429 (1996 Supplementary Pamphlet); 7A U.L.A. 1 (1996 Cumulative Annual Pocket Part).

the FTC's proscription of non-purchase money security interests only when the collateral is household goods, the UCCC generally prohibits the use of any non-purchase money collateral to secure credit sales. With exceptions permitting certain security interests in goods in which purchase money collateral is installed and land to which purchase money collateral is affixed, a seller, and any assignee of the seller's right to payment,[100] is limited to a security interest in the property sold.[101]

That purchase money collateral is permitted to secure indebtedness arising from prior and subsequent secured credit sales as well as the indebtedness arising from its sale.[102] The benefit to the seller of such cross collateralization is considerably diminished, however, by a UCCC provision that applies payments first to debts arising from the sales first made and terminates security interests in property as the debts arising from the sale of that property are paid.[103] In the FTC's rejection of a similar provision limiting the benefit of cross collateral, the FTC opined:

> This provision could significantly reduce the value of purchase money security interests. No payments would be allocated to reducing the principal owed on the most recent purchase until all earlier purchases are paid off. If preceding purchases were not paid off until a year after the most recent one, for example, the only security for the entire amount of the credit extended for the most recent purchase would then be a year-old appliance or piece of furniture.[104]

In addition to prohibiting the taking of virtually all non-purchase money security interests by sellers and their assignees, and limiting the amount of indebtedness that may be secured by the use of their purchase money security interests as cross collateral, the UCCC abolishes a basic remedy of secured credit for those security interests permitted those creditors. In this instance, the UCCC rule applies not only to sellers and the banks and finance companies who are assignees of their rights to payment but also to certain direct lenders to consumers whose close working relationship with sellers is not unlike that of the assignee, indirect financer.[105] The UCCC imposes upon these purchase money secured creditors an election of remedies when the price arising from the sale of the goods to the debtor is $1,750 or less. They have the option of suing for the unpaid balance of their claims or repossessing, but they may not do both.

---

100. U.C.C.C. § 1.301 (38) (1974).
101. *See id.* at § 3.301.
102. *See id.* at § 3.302.
103. *See id.* at § 3.303.
104. FTC Rule, *supra* note 50 at 7786.
105. *See* U.C.C.C. § 3.405(1) (1974).

The provision does contain exceptions that allow the creditor to hold the consumer liable if the consumer has wrongfully damaged the collateral or if, after default and demand, the consumer has wrongfully failed to make the collateral available to the creditor.[106]

The Comment to the UCCC provision invoking the election of remedies doctrine contains no justification for the rule, but plainly, the draftspersons must have applied the lost-value premise to collateral that in their opinion would probably not have as high a resale value to the secured creditor as its retention value to the debtor. The FTC considered but rejected an election of remedies doctrine for purchase money security interests in household goods, concluding that the provision could raise the costs and reduce the availability of credit in excess of offsetting benefits.[107]

A final incursion by the UCCC on rights traditionally associated with secured credit—this one applicable to direct lenders to consumers as opposed to credit sellers and the assignees of their rights to payment—voids in some instances the rule that exemption laws do not prevent a secured creditor's repossession and forced sale of collateral.[108] The provision applies to collateral for a supervised loan, which is defined as "a consumer loan...in which the rate of the finance charge, calculated according to the actuarial method exceeds 18 per cent per year."[109] The provision is further limited to non-purchase money security interests in property other than motor vehicles. It protects only exempt property that is being used by the debtor or a member of a family wholly or partly supported by the debtor. When these conditions are met, the secured creditor may not repossess the collateral without a court order, which may be issued only after a hearing following notice to the debtor. The court may not authorize repossession of the collateral "if it finds upon the hearing both that the consumer lacks the means to pay all or part of the debt secured and that continued possession and use of the [collateral] is necessary to avoid undue hardship for the consumer or a member of a family wholly or partly supported by him."[110]

This provision of the UCCC limited to certain lenders may be compared with the FTC's proscription of non-purchase money security interests in household goods by all sellers and lenders. For affected lenders, the UCCC provision is broader in scope as it covers all exempt property except motor vehicles, and while the provision does not expressly pro-

---

106. *See id.* at § 5.103.
107. *See* FTC Rule, *supra* note 50 at 7784.
108. U.C.C.C. § 5.116 (1974).
109. *Id.* at § 1.301(43).
110. *Id.* at 5.116.

hibit security interests in exempt property, it significantly undermines the value of such security interests. The criterion for denying the secured creditor repossession of the collateral at the hearing is, not one that weights the debtor's needs against the resale value of the collateral to the creditor, but one that looks solely to the debtor's needs. And even if the creditor prevails at the hearing, the requirement of a hearing and judicial order prior to repossession imposes costs that self-help repossession or judicial process providing for prejudgment seizure, common remedies of secured creditors, do not.

Although adoption of the UCCC's provisions would significantly increase the protection given defaulting debtors in most states, Jeffrey Davis, now my colleague at the University of Florida, writing in 1973, reported that a major obstacle to the passage of the 1968 UCCC by more states in its early years was a lack of general support by consumer groups. He observed that initial consumer approval of the UCCC changed to mounting criticism that "centered around an alleged failure to restrict creditor powers and practices sufficiently."[111] Apparently this consumer-group dissatisfaction with the UCCC has continued, while coupled no doubt with the efforts of creditors to avoid further regulation, for the seven states that had enacted the UCCC in some form by 1973[112] increased by only four to 11 in 1996.

Davis reported that the most significant anti-UCCC action was taken by the National Consumer Law Center, which, acting on the consensus of some fifty-five consumer experts that the UCCC was inadequate and obtaining no input from creditors, drafted the National Consumer Act (NCA), published in 1970. Davis relates that the drafting committee consisted solely of avowed consumer advocates, that predictably the NCA extended well beyond the UCCC in restricting creditors' remedies, and that even its supporters admitted that it was often guilty of overkill.[113] While neither the NCA or its successor, the Model Consumer Credit Act, which restricts the taking of even purchase money security interests in some instances, have been enacted in any state in their entirety, provisions of these model acts have influenced the regulation of creditors' remedies in some states.[114]

---

111. Jeffrey Davis, *Legislative Restriction of Creditor Powers and Remedies: A Case Study of the Negotiation and Drafting of the Wisconsin Consumer Act*, 72 MICH L. REV.1, 4 (1973) (citation omitted).
112. *Id.* at 5, citing 1 CCH CONSUMER CREDIT GUIDE ¶ 4770 (1973).
113. *Id.* at 4.
114. 1 HOWARD J. ALPERIN & RONALD F. CHASE, CONSUMER LAW § 105 (1986). The Model Consumer Credit Act restricts the taking of purchase money security interests in "personal effects, household furnishings, appliances and clothing of the consumer and

Analysis of more existent or proposed proscriptions on supplemental remedies will only provide further evidence of what should now be all too obvious: the number of proscriptions that may be imposed equal the number of supplemental remedies. Moreover, the same rationale used to ban supplemental remedies—the lost value premise—could be used in many instances to attack the basic postjudgment execution or garnishment process, although the procedural safeguards and other protections, most notably exemptions, built into that process, and the need to maintain some means of enforcing the law of obligation, have served thus far to preserve postjudgment judicial process from attack.

The flurry of activity that marked state and federal action regulating the use of supplemental remedies in the latter 1960's and in the 70's and 80's would seem to have subsided somewhat in this decade, but the ebb of regulatory action may again turn to flow in the waning years of this millennium or the early ones of the next. Hence, further analysis of the lost-value premise, combining broader perspectives on the values at stake in that premise with consideration of the enhanced means that now exist for preventing coercive collection that may, at the time it is exercised at least, cost the debtor more than it benefits the creditor, is more than an academic exercise.

A second reason for the timeliness of the analysis of the lost-value premise that follows is that it shows the need for creditors in some instances to take the offensive in "remedies' wars," primarily by changing exemption laws, often ones of long-standing. By unjustifiably restricting the effectiveness of the basic judicial collection process, some exemption laws have contributed significantly to creditors' need for the now besieged supplemental remedies.

The principle upon which exemption laws are based is again, I believe, the lost-value premise. At the time a debtor's last assets are taken, the loss of marginal utility to the debtor may well exceed any marginal benefits to creditors or, by extension, to debtors who pay their obligations. And so long as exemption laws are limited to property that the debtor needs to maintain economic viability—even, most of us would readily concede, a certain minimal comfort level—they are justifiable even when subjected to the expanded analysis of lost value that considers the default-deterring benefit of the debtor's commitment to payment that stems

---

his dependents." MODEL CONSUMER CREDIT ACT § 2.411(2)(b) (National Consumer Law Center 1973). In Davis's report of the background, negotiation, and drafting of restrictions on creditors' remedies in the Wisconsin Consumer Act, he notes that, while that Act is based on the National Consumer Act, "the substantive changes that resulted from negotiation and redrafting are so pervasive that its resemblance to the NCA is at best slight." Davis, *supra* note 111 at 5, n.17.

from the potential for lost value. Like other lost-value justifications for restricting creditors' remedies, the justification for exemptions recognizes that the credit industry and not taxpayers should bear the cost of providing the economic safety net necessitated by the marketing of credit. But exemption laws in some states go far beyond the point of maintaining the debtor's economic viability, far beyond their justification on an expanded lost-value rationale. This occurs most flagrantly when exemptions are provided for particular types of property, and no limit is placed on the market value of that property.

My state of residency provides striking examples. In Florida the homestead exemption immunizes 160 acres of realty located outside a municipality or one-half acre of land within a municipality from the claims of creditors who do not have a statutory lien for improvements on the homestead or a mortgage.[115] Within these parameters, the debtor's home and the land upon which it is situated may be worth hundreds of thousands or even millions of dollars, but a creditor with an unsecured claim against the debtor will be powerless to enforce a judgment on that claim against the homestead property. Nor does Florida place monetary limits on its exemption of the cash surrender value of life insurance, the proceeds of annuity contracts[116] and wages, except by waiver of exemption that may run afoul of the FTC's-Credit Practice Rule, when they are the earnings of a head of family.[117]

California and Texas as well as Florida are known for their liberal exemption laws,[118] but examples of exemptions that exceed reasonable policy justifications surface not infrequently in cases in which the exemptions are provided by states without a reputation for largess for debtors facing judicial process. For example, the Oklahoma exemption for "implements of husbandry necessary to farm the homestead" that came before the U. S. Tenth Circuit Court of Appeals in a 1986 bankruptcy case contained no maximum-value limitation and thus exempted a $30,000 tractor.[119] And in a 1988 Eighth Circuit bankruptcy case, a debtor had liquidated some $700,000 worth of non-exempt property and invested the proceeds in life insurance and annuity contracts with a fraternal benefit association, which property was then exempt without regard to monetary limit under Minnesota law.[120]

Statutes that permit debtors to immunize property of unlimited value from the claims of their creditors are a mockery of our law of civil oblig-

---

115. FLA. CONST. art. X, § 4.
116. FLA. STAT. § 222.14 (1995).
117. *See id.* at § 222.11.
118. *See* WARREN & WESTBROOK, *supra* note 82, at 136–37, 154–55.
119. *In re* Liming, 797 F.2d 895 (10th Cir. 1986).
120. Norwest Bank Nebraska v. Tveten, 848 F.2d 871 (8th Cir. 1988).

ation. To the extent that a debtor exempts a particular type of property in an amount that appreciable exceeds that commonly held by similarly situated debtors who are not contemplating bankruptcy, exemption laws favor debtors who have the resources, in spite of their insolvency, to engage in bankruptcy planning. Thus, unlimited exemptions undermine principles of fairness in addition to unduly penalizing debtors who pay.

Some measure of protection for creditors is provided by general fraudulent transfer law and recently enacted statutes that expressly empower creditors to levy upon otherwise exempt property that the debtor has obtained by fraudulent conversion of non-exempt assets.[121] These provisions, however, are less effective than placing reasonable value limits upon all categories of exempt property or assigning the debtor a maximum exemption value without restricting the types of property it could be used to protect.

In the past, overly generous exemptions may have been fueled by something more than populist sentiment or special-interest lobbying if they were designed to protect debtors against precipitate actions of their creditors. But if that justification ever warranted these exemptions, certainly the debtor's now increased ability to stem coercive action by creditors through the use of counselor-sponsored workouts adds appreciable weight to the argument that these exemptions should be modified.

(4) Reevaluating the Lost-Value Premise as a Basis for Proscribing Creditors' Remedies

If one finds that the exercise of certain remedies thrust losses on debtors that do not result in corresponding recoveries for their creditors, there are two reasons for concluding that this concept of lost value alone may not justify the proscription of those remedies. One is Scott's argument that the remedies may be justified when the credit transaction is viewed from the predefault perspective, where they reduce the costs and increase the availability of credit by providing incentives for performance of the contract by the debtor. The other reason is that advanced in this study. Consumer counseling agencies curtail the counterproductive use of remedies by correcting misdiagnoses of certain collection cases as zero-sum games and obtaining the cooperation of creditors in all cases that qualify as ones of Creditor's Dilemma.

These two reasons for preserving creditors' remedies against attack based on the postdefault lost-value premise are not in conflict. Scott's argument that value-destroying remedies furnish an incentive for the debtor

---

121. *See e.g.* FLA. STAT. § 222.30 (1995).

to perform the contract is not undermined by recognizing that a qualified debtor may also escape the damages these remedies may inflict by a counselor-assisted workout. Once timely payment is no longer possible but workout is, the possibility of creditors' use of remedies that may destroy value motivates the debtor to pursue the best action for creditors that her changed circumstances permit—a counselor-assisted workout. And so long as any remedy available to a creditor has primarily deterrent as opposed to pecuniary value, that creditor's cooperation in the workout will be easier to obtain.

Prior to common access by debtors to Consumer Credit Counseling agencies, beneficial workouts often failed to materialize due to creditors' doubts as to the debtor's ability or resolve, or the inability of any one creditor to insure that if he granted extensions others would not step in front of him in the collection line by continuing to apply coercive measures. In these instances, debtors could escape creditors' coercive actions that might produce lost value only by succeeding in the often difficult job of securing creditor cooperation by self-advocacy, filing for bankruptcy, secretion of assets or clandestine changes of residence. These alternatives, however, placed other costs on both debtors and their creditors that need not have been incurred in instances in which workouts were feasible. The qualified debtor who seeks the intervention of a Consumer Credit Counseling agency and who adheres to the plan the agency sponsors may now avoid the costs of any of these alternatives to coercive collection or of coercive collection itself.

If the debtor's financial burden is beyond the scope of the four-year workouts sponsored by counseling agencies, or if the debtor's resolve is lacking, either Chapter 7 or Chapter 13 bankruptcy may be invoked to foreclose the use of coercive creditors' remedies that may otherwise produce lost value. While bankruptcy imposes its own kinds of pecuniary and psychic costs on the debtor, preservation of the value of assets has always been a principal goal of bankruptcy law. Thus, even the debtor who cannot or will not use the services of Consumer Credit Counseling agencies carries the key to avoiding lost value from coercive creditors' remedies. The important change added by a readily available credit-counseling alternative in suitable cases is that avoidance of lost value no longer systematically requires that the debtor who does not hide himself or his assets incur the costs, including the stigma, associated with the filing of a bankruptcy petition. Even that rarity, the Chapter 13 case in which the debtor pays all claims in full, will stigmatize the debtor in ways that may be avoided in a non-bankruptcy workout, perhaps because so many Chapter 13 cases result in so little payment to unsecured creditors that many people fail to distinguish those that do.

Admittedly, the services of consumer credit counseling agencies are not reasonably accessible to debtors in all areas of this country. But there is no reason to believe that these agencies will not continue to move into new areas, diminishing the problem of inaccessibility as they do. Debtors living in more sparsely populated areas may reasonably be expected to suffer greater inconvenience in making use of counseling agencies. Moreover, in these areas more traditional patterns of relationships contribute to creditors knowing one another and their debtors better than they do in cities. In small communities and rural areas, where counseling agencies are less likely to be available, promising workouts may be more easily implemented without the intervention of those agencies.

When the debtor's enhanced ability to substitute workouts for coercive collection is given due recognition, even that myopic assessment of lost value that single-mindedly focuses on differentials at the time property is taken loses much of its law-reform appeal. The argument in favor of the proscription of remedies based on this narrow concept of lost value proves too much, for all remedies, even postjudgment judicial ones, are capable of producing greater loss to the debtor than gain to the creditor at the time they are exercised.

As observed earlier, a more comprehensive test of lost value asks whether any difference in the value of property at the time it is taken by exercise of a suspect remedy is offset by the value of that remedy as a deterrent or curer of default in the many instances in which default either does not occur or is timely cured. Certainly this more comprehensive test should be the one that commands the attention of lawmakers bent on providing a legal environment in which market forces will provide an adequate supply of credit at affordable rates.

The FTC recognized the more comprehensive test of lost value in its analysis of reasons for promulgating the Credit Practices Rule and certainly professed to apply that test in its numerous manifestations of concern with the effects of the rule on the market for consumer credit. Unfortunately, as the FTC's analysis reveals, the effects of the loss of various remedies on the cost and availability of credit are difficult to measure. In my analysis of the Credit Practices Rule in the preceding section of this chapter, I concluded that the FTC had not proved that costs outweighed benefits to consumers for the remedies it proscribed. The FTC's debatable determination of that issue, however, probably manifested more concern with applying the more comprehensive test of lost value than many actions of state legislatures.

In a market economy, whether creditors' contract with their debtors for supplemental remedies, like other practices in the distribution of goods and services, would normally be left to market forces. Parties to a credit contract could then assess the net costs or benefits of supplemental remedies and tailor their contracts accordingly. The FTC recognized this principle in its analysis of the Credit Practices Rule but found that market imperfections prevented the forces of supply and demand from maximizing benefits and minimizing costs in the instances of the remedies it banned.[122] Because the FTC's justification for the rule rests upon its finding of market failure, consider in full the case for that failure put by the FTC:

> In consumer credit transactions, the rights and duties of the parties are defined by standard-form contracts, over most of which there is no bargaining. The economic exigencies of extending credit to large number of consumers each day make standardization a necessity. The issue, however, is whether the contents of these standard form contracts are a product of market forces.
>
> Although market forces undoubtedly influence the remedies included in standard form contracts, several factors indicate that competition will not necessarily produce optimal contracts. Consumers have limited incentives to search out better remedial provisions in credit contracts. The substantive similarities of contracts from different creditors mean that search is less likely to reveal a different alternative. Because remedies are relevant only in the event of default, and default is relatively infrequent, consumers reasonably concentrate their search on such factors as interest rates and payment terms. Searching for credit contracts is also difficult because contracts are written in obscure technical language, do not use standardized terminology, and may not be provided before the transaction is consummated. Individual creditors have little incentive to provide better terms and explain their benefits to consumers, because a costly education effort would be required with all creditors sharing the benefits. Moreover, such a campaign might differentially attract relatively high risk borrowers.[123]

The FTC's findings may be attacked on two grounds, the first critical of the remedy chosen by the agency and the second questioning whether the paucity of contracts devoid of remedy terms offensive to the FTC that existed in at least some sectors of the market was properly attributable to market failure.

Each factor that the FTC cites in support of its guarded conclusion that "competition will not necessarily produce optimal contracts" is one that readily lends itself to correction by rules providing for meaningful pre-contractual disclosure of remedies by a statement in plain language containing a brief description of applicable remedies. Instead of dictating the exclusion of contractual terms, an extreme measure when tested against the standard of our free-market model, much of the thrust of consumer protection law in this country has been directed toward respecting

---

122. FTC Rule *supra* note 50 at 7744.
123. *Id.*

consumer choice while trying to insure that that choice is an informed one.

The similarities between a consumer shopping for better remedy terms and one shopping for better interest rates, before the passage of Truth in Lending standardized credit-cost disclosure, did not escape the attention of the FTC. It concluded, however, that while adequate information about interest rate options are now available to consumers, "it does not follow that in other credit areas efficient options will be made available since there may be other market failures, which prevent suppliers from offering them."[124] The FTC's analysis of the rule does not offer, however, any causes of market failure other than those contained in the above citation,[125] which appear to lend themselves readily to cure by adequate disclosure provisions.

Criticism of the FTC's method of correcting market failure is overshadowed, however, by a more fundamental question. Was the FTC's finding of market failure, which is recognized by the FTC, as it should be by any lawmaker, as a condition precedent to governmental edict encroaching upon freedom of contract,[126] justified?

In the context of contract terms providing for supplemental creditors' remedies, a determination of market failure requires that lawmakers find that informed consumers would prefer removal of such terms to the consumer benefits that Scott's analysis of deterrent value demonstrates they may provide—paying lower interest rates or obtaining credit that would otherwise be unobtainable. The validity of such a finding by lawmakers is undermined not only by recognition of these predefault benefits of supplemental remedies, which may more than offset postdefault costs, but also by recognition of consumers' enhanced abilities to avoid or reduce those postdefault costs by greater access to workouts, a cost-mitigating factor that the FTC nor any other lawmaker, of which I am aware, has yet taken into account. Thus, it is highly unlikely that consumers would enter into contracts containing supplemental-remedy terms only because they have no other choice or, if they have such choice, exercising it in favor of contracts containing supplemental remedy terms would constitute cognitive error on their part.

Scott observes that related to a justification that proscribes certain creditors' remedies based on debtors' inability to properly assess potential damages to themselves is a justification that prohibits those remedies

---

124. *Id.* at 7788.

125. The FTC's compendium of the causes of market failure contained in the citation in the text at note 123 is a second account of these causes in different language than the first. *See id.* at 7746–47.

126. *See id.* at 7744–45.

because of the damages that may inflict on society. These externalities result from a pervasive system of social-insurance programs that prevent debtors from internalizing fully the social costs of default.[127] Therefore limitations on remedies may "be a crude but presumably effective method of credit rationing."[128] Scott rejects, however, this second aspect of the market-imperfections justification for proscribing creditors' remedies by noting that curbs on coercive remedies will exacerbate third-party effects by encouraging "risk seeking behavior by those who obtain installment credit."[129]

But if the market-imperfection rationale for proscription of remedies is flawed and not the consumer credit market, what other factors may have motivated lawmakers to attack supplemental remedies? What other values are at stake in regulating supplemental remedies?

Concern for the inability of less affluent, less financially sophisticated consumers to order their financial affairs appears to have significantly influenced the FTC's actions in promulgating the Credit Practices Rule. The case for an underlying paternalistic motive for proscribing remedies in this instance finds support in a comparison of the previous use of remedies that were banned in the Credit Practices Rule with the use of those remedies that escaped that action, although they had been candidates for proscription in the rule the FTC had proposed earlier.

Terms providing for confessions of judgment, waivers of exemptions, blanket security interests in household goods and wage assignments in states where one or more of these remedies were permitted prior to the Credit Practices Rule were more likely to be found in contracts used in extending credit to less affluent consumers than middle-class ones. Although the evidence is incomplete and ambiguous in some respects, it would appear that contracts providing for these remedies were much more frequently used by small loan companies, finance companies, and credit unions and banks when catering to the needs of those with less income to repay their obligations.[130] And if middle-class debtors' credit contracts

---

127. *See* Scott, *supra* note 1 at 771–72.
128. *Id.* at 771
129. *Id.* at 772.
130. FTC Rule, *supra* note 50 at 7746 ("Finance companies in particular are more likely to use the remedies subject to this rule than are other creditors."), 7757 ("The rulemaking record shows that wage assignments are used primarily by small loan and finance companies."), 7762 ("Based on the rulemaking record, we find that the practice of securing consumer loans with a non purchase money security interest in household goods (HHG) is widespread. Finance companies are the preeminent users, and HHG security interests are found in a majority of finance company loan contracts. However, banks also avail themselves of such security as do credit unions and even occasionally,

did contain one or more of the now federally prohibited terms, those debtors were far more apt to seek the services of attorneys who could threaten or file bankruptcy proceedings or credit counselors who could get them leeway for workouts. By these means or by effective self-advocacy of their ability to effect workouts, middle-class debtors could better shield themselves from the use or undermining threat of use of a punishing remedy than most financially unsophisticated and less affluent debtors.

But the rejected proposals in the FTC rule—valuing collateral other than household goods at its retail price for purposes of calculating deficiencies, election of remedies in the case of purchase money security interests in household goods, and reduction of the benefits provided by cross collateralization in installment sales—would have restricted practices that apply across the entire range of consumer lending. Providing consumers protection from these practices would have exceeded the limits of any reasonable argument based on paternalism, for it would have imposed costs on the great majority of consumers for protection that they are quite capable of providing themselves.

If the proscription of some remedies that impact primarily on those least able to protect themselves is to be justified on paternalistic grounds, however, that justification must be balanced with the special needs of creditors who make loans that more conservative creditors will not. Unless high-risk lenders can keep their losses within acceptable bounds, an insufficient supply of credit from legitimate lenders will exacerbate the economic problems of many low-income consumers.

---

savings and loan associations. Although retail installment sales acts tend to restrict retailers to a purchase money lien on the goods sold, the record also reveals that certain retailers rely on HHG security interests as additional collateral in credit sale transactions.").

Evidence in the FTC rule, however, did not reflect that finance companies used confession of judgment or cognovit clauses more than other lenders.

> A survey of its members conducted by the Consumer Bankers Association... shows that approximately 20 percent of banks responding to the survey included cognovit clauses in the majority of its contracts where permitted by law. A survey of legal aid attorneys indicates that, where permitted by law, cognovit clauses were utilized in 20 percent of loan agreements by credit unions, 21 percent by finance companies, 16 percent by banks, and 30 percent by creditors generally....[A] 1970 industry survey conducted by the National Commission on Consumer Finance showed that 17 percent of large bank respondents and 17 percent of large finance companies stated cognovits to be a highly valuable provision in contracts for unsecured cash loans. This suggests that, among these respondents, confessions of judgment are employed on a regular basis. *Id.* at 7752 (citations omitted).

The FTC rule gives no evidence of the use of waivers of exemption by different types of lenders. *See id.* at 7769.

Those who give most credence to paternalistic concerns, however, may perceive credit rationing as a benefit to less sophisticated consumers. The dynamics of selling, even selling goods that are objectively what they purport to be in transactions that are free of fraud if not of hype, differ in many respects not so much in form as in degree from those of swindling, and both seller and swindler are advantaged by transactions in which their take is not limited to the cash that the buyer or mark has on hand at the time the deal is made.[131] Credit then compounds the harm that buyers can inflict on themselves by the mistakes they make in the marketplace.

But credit rationing fails to distinguish the many times the consumer uses credit advantageously. And the sole determiner of when credit is used advantageously in a free society must necessarily be the consumer. Avoiding the pressures of paying for extravagances is something we must all learn for ourselves and in our own way. Maybe along that way we may be so fortunate as to learn that happiness comes far less from possessions than from realizing our interconnectedness with others, with nature and, when we are attuned, perhaps even the power behind nature and others. But, alas, credit rationing, like other forms of paternalism, is indiscriminate. It fails both to recognize our need for necessities and at least a modicum of the frills of the marketplace and our need to arrive at higher, less material values by our own volition and in our own way.

Paternalism obviously does justify the promulgation of some rules to regulate the credit market. Few would champion the enforcement of contracts of indentured servitude by specific performance, and even when obligations are purely monetary, few would deny that overwhelming obligations, at least if they are not fraudulently incurred or the result of other willful wrongs, should not be dischargeable in bankruptcy. But the creditors' supplemental remedies that have been the subject of this chapter do not, as Scott has observed, sacrifice the debtor's personal integrity. Nor does the paradox of rational decision makers acting on inconsistent preferences—ignoring long- term preferences in order to indulge short-term desires—necessarily justify paternalistic regulation.[132]

Scott examines a further reason that may explain if not justify the current scheme of regulation of supplemental creditors' remedies: "lost value may be a proxy for distributional equity."[133] Drawing upon the work of Yale Law School Dean Anthony Kronman, Scott states that "[a] redistributive justification for regulating self-enforcing terms rests on the premise that it is 'fairer' or more just to change the current distribution

---

131. *See generally* ARTHUR ALLEN LEFF, SWINDLING AND SELLING (1976).
132. *See* Scott, *supra* note 1 at 776–82.
133. *Id.* at 787.

of rights upon default in a way that helps one party or group at the expense of another."[134] Prohibiting self-enforcing remedies would, at least on initial examination, shift rights from creditors to debtors.[135] But not all power imbalances are cause for social concern, and in an effort to determine whether power should be transferred to debtors in this instance, Scott applies William Baumal's "superfairness" theory.[136]

The process of distributional equity under end-state superfairness is designed to insure that after the change in distribution of rights, each class of participants prefers its own share or is indifferent to that share and a share received by another class.[137] Obviously the class from whom rights are taken will not be indifferent to the taking, but that is of no concern to end-state superfairness, which looks solely to absence of envy among the classes of participants following the transfer. While the prohibition of creditors' remedies does redistribute postdefault leverage from creditors to debtors, distributional equity is undermined by the recognition that debtors who do not default will likely bear the burden in higher interest charges of those who do.[138]

Scott finds that the appeal of distributional equity may be found, however, in the "economic illusion" that weights concentrated losses, such as those that are inflicted on a debtor whose property is seized, more heavily than diffused gains, such as avoidance of the higher charges that would be incurred by all debtors who pay to cover losses associated with prohibitions on various creditors' remedies.[139] Scott states that "the illusion that causes concentrated losses to overwhelm diffused gains may cause each affected individual to prefer her end-state distribution to that of any other individual."[140] He concludes that "[t]he idea that policy is mo-

---

134. *Id.* at 782 (*citing to* Anthony T. Kronman, *Paternalism and the Law of Contracts*, 92 YALE L.J. 763, 770–74 (1983); Kronman, *Contract Law and Distributive Justice*, 89 YALE L.J. 472, 499–501 (1980)).
135. *See id.*
136. *See id.* at 783.
137. *Id.* at 783.
138. *See id.* at 784–85. Scott supports his conclusion that debtors who pay and not creditors will likely bear the burden of increased bad-debt losses due to proscription of supplemental remedies with the following microeconomic analysis:
> If supply is elastic and demand is not (along the relevant range), creditors will be able to raise the price of credit to reflect the distributional change, and thus the burden of the regulation will rest largely on those debtors who pay.... The common assumption about capital markets is that barriers to entry and exit are low, implying that supply is elastic. Thus, it is likely that at least some of the burden of the regulation will fall on debtors who pay. *Id.* at 784 (citation omitted).

139. *See id.* at 785–86.
140. *Id.* at 788.

tivated by an aversion to concentrated losses does offer the most plausible positive account of the regulation of self-enforcing remedies."[141]

If the FTC was motivated by the economic illusion that causes concentrated losses to overwhelm diffused gains, that motivation, like my belief that paternalistic concerns motivated the FTC, must be gleaned from sources other than the FTC's stated reliance on market imperfections as its justification for the Credit Practices Rule. Evidence of the FTC's concern with avoidance of concentrated losses, even though that avoidance entails costs to consumers who do not default, can be found, however, in its analysis of the injury to consumers in the remedies it banned.

> In assessing particular remedies, our focus has been on the consequences of this remedies [sic] for consumers in those cases when the remedy is invoked or threatened. Nonetheless, all consumers will benefit from the rule to the extent that it reduces the adverse consequences of default because it serves, in that capacity, as a form of insurance. At the time a consumer enters into a loan agreement, the likelihood of default is both remote and difficult to assess. Thus all consumers face some risk of default and will value insurance, which reduces the most injurious consequences of default, even if they never need the insurance. In this sense, all consumer debtors will benefit.[142]

In a footnote the FTC observes that "[t]he insurance thus provided is not costless, of course, and some consumers may prefer not to purchase it."[143]

Scott's discovery, after lengthy search, of a basis for the appeal to lawmakers of regulating supplemental remedies does not mean, however, that he accepted that basis as a justification for the law it has spawned. He views the illusion-of-equity premise upon which the distributive justice case for regulation is based as a competing normative theory with his own, and one which willingly sacrifices the predefault deterrent value upon which his justification for supplemental remedies is based.[144] Scott concludes that the selective regulation of particular supplemental remedies will lead to pressure for substitute enforcement mechanisms unless some means is found to "ameliorate the information deficits that generate coercive action."[145]

In his conclusion then, Scott solidifies his position in one important respect with the commentators on the subject of lost value who preceded him. From Leff's pioneering work ascribing the source of our dissatisfaction with supplemental remedies to lost value occurring at the time the debtor's property is taken, to Scott's more comprehensive analysis

---

141. *Id.*
142. FTC Rule, *supra* note 50 at 7744.
143. *Id.* at 7744, n.25.
144. Scott, *supra* note 1 at 785, n.190.
145. *Id.* at 788.

finding an overriding net benefit in these remedies by valuing the deterrents to default they provide, one aspect of the ongoing analysis has remained constant. Both Leff and Scott viewed asymmetries of information between the debtor and creditor as the principal reason for the failure to substitute cooperative collection for value-destroying coercive collection. And writing in the interim between the works of these two commentators, Whitford expanded the scope of the informational deficits that impede cooperative recovery efforts. He observed that cooperative collection may require something more than that the debtor and her aggrieved creditor possess accurate information about the capabilities and intentions of each other. What is also needed in many cases of serious delinquency is some means of terminating the race of diligence among creditors. That requires that each creditor know that the debtor's other creditors will cooperate in a workout. Without some means of monitoring other creditors' conduct, the cooperating creditor assumes the risk that they may accelerate their recovery by continuing to use coercive measures, which shift more of the risks of an ever riskier workout to the cooperating creditor.

In examining the dynamics of the collection process, this work too has found that the principal cause of the counterproductive use of supplemental remedies is informational deficits. This study has also revealed, however, something that has gone hitherto unheralded in the published deliberations of lawmakers and analyses of commentators: the emergence of an agency, Consumer Credit Counseling Services, that bridges the informational gaps in collection cases and therefore provides a cost-effective means of obtaining creditor cooperation in cases of promising workouts. Now widely available to debtors in this country, these counseling services provide a cost-effective means of preparing workout plans when they are feasible, procuring creditor acquiescence in extending payment terms, and monitoring the actions of the debtor and creditors during the workout. Consumer Credit Counseling Services offer a needed alternative to overextended debtors who otherwise might suffer the pangs of an unsuccessful unassisted workout, the stigma of an unwanted bankruptcy, or both.

Avoiding costs to debtors that are not perceived as resulting in corresponding gains to their creditors and therefore avoiding lost value has long been the principal argument advanced for proscribing supplemental creditors' remedies. Scott's analysis significantly diminishes the superficial appeal of this argument. He finds that a comparison of the value of property to debtor and creditor at the time a supplemental remedy is used is an incomplete analysis of the utility of the remedy, and that consideration must also be given the important default-deterring function of supplemental remedies. Recognition of the role that Consumer Credit Counseling agencies play in providing the debtor a means of preventing the use of supplemental remedies with lost-value potential provides a second reason for rejecting the attack on these remedies. Overextended debtors now have the power to block counterproductive collection practices in all the various stages of overextension in which they may find themselves. They may use some form of bankruptcy when payment of all claims in full over an extended time is not feasible and counselor-assisted workouts when it is.

For debtors not paying their obligations who fall in neither category because they are able with adjustments in budget or otherwise to make timely payment, creditors need effective remedies to reduce the costs of credit and clearly signal that Americans respect the honoring of commitment. Adequate provision by government for the enforcement of monetary obligation is obviously important for the amount of recoveries that are effected by the use of remedies that the state provides directly or sanctions as enforceable supplemental remedies provided by credit contracts. But governmental support for the enforcement of obligation serves a larger purpose. It reinforces that basic principle upon which all human cooperation that is not purely altruistic is based: we may expend our resources for others because we may rely on their promises of remuneration.

We cannot afford to erode this principle by condoning signals from lawmakers that obligation is not to be taken seriously. Reciprocity drives market economies, and it is simply wrongheaded, in the absence of some strong competing concern, to convert what were intended as mutual exchanges into involuntary takings by the loss of cost-effective remedies. Recognizing that the majority of overextensions are primarily due to nonvolitional expenses or interrupted income does not militate against providing creditors with effective remedies to use against debtors who fail to make payment when they are able to do so. Nor does enforcement of market obligations further only materialistic values. While pursuit of the frills of the marketplace, which also accounts for much overextension, may blind us to higher cultural, intellectual and spiritual values, loss of respect for commitment to market obligation may erode these higher values as well as the fundamental tenet upon which our credit economy is based.

C. Preserving the Unique Role of Consumer Credit Counseling Agencies

Previous examination of the work of counseling agencies has revealed that these agencies make possible the debtor's financial rehabilitation by securing the voluntary cooperation of creditors. Unlike a debtor in a Chapter 13 bankruptcy proceeding, however, there is no automatic stay that forces creditors to forego their individual collection efforts when a credit counselor proposes a workout.[146] While counseling agencies affiliated with the National Foundation for Consumer Credit have generally been effective in securing the cooperation of creditors, the agencies' efforts in that regard are not always successful and, although cost-effective, are never costless. Would the work of the agencies benefit from a law that limited an individual creditor's actions to collect once a counseling agency sponsored a debtor's workout, even though that creditor had not acquiesced in the workout?

Such law need not be as broad in its proscription of creditors' judicial and self-help remedies as the automatic stay in bankruptcy proceedings. The Ohio procedure for issuance of an order of garnishment of nonexempt personal earnings is illustrative. Before such a garnishment order may be issued, the creditor must serve a written demand on the debtor.[147] One of the methods by which the debtor may avoid garnishment of personal earnings is by entering into an agreement with a credit counseling service for "debt scheduling."

The obvious benefit of the Ohio statute is that creditors who threaten wage garnishment in an effort to secure advantages over other creditors are precluded from imposing this obstacle to a plan that may further the collective good of the debtor and all her creditors. Still, there is good reason for legislatures to proceed with caution in imposing non-volitional requirements on creditors whose debtor has undertaken a counselor-assisted workout.

Debt adjustment through consumer credit counseling agencies is the only systematic method of addressing the issue of serious overextensions by consumers that does not rely on the power of the state in the form of either the use or threat of use of creditors' remedies or bankruptcy. I believe it is important to preserve unabated this fundamental difference in methods of resolving consumers' financial troubles. That which is founded on agreement, as are obligations incurred in the marketplace, are often best adapted to changed conditions by further agreement, provided the party in default continues her good-faith efforts. The alternative is often a futile and costly attempt to obtain the unobtainable.

The voluntary nature of counselor-assisted debt adjustment serves as a check on credit counselors' sponsoring too many cases with too low a probability of success. Moreover, the voluntary nature of participation in these workouts leaves debtors and creditors with a feeling that they

---

146. *See* 11 U.S.C.A. § 362 (West 1993 & Supp. 1996).
147. *See* OHIO REV. CODE ANN. §§ 2716.02, 2716.03 (A)(3) (Page Supp. 1996).

are still in control of their financial affairs. If garnishment of non-exempt personal earnings is a creditor's only effective remedy and that remedy is lost when the debtor enters into an agreement for a workout with a credit counseling service, then that creditor is as much a captive of the counseling service's proposed workout as he would be of a plan confirmed by a bankruptcy court.

Painting bright lines around the graduated responses available to financially troubled debtors furthers a broader purpose. When the law of collecting obligations is compared with the law establishing them, most observers would probably conclude that the state gives creditors considerably less assistance in collecting obligations than in establishing them. If the law teaches that debtors are not required to do everything possible to further payment of their obligations incurred in the marketplace—and the additional remedies given creditors collecting support obligations in domestic relations cases show that the state can fashion more potent remedies—what do Americans reasonably expect of debtors who are overextended on market debt? I know of no empirical data that answers that question, but impressionistic evidence suggests that a debtor is expected to pick from the graduated responses available to him that response that affords him no greater concessions than his circumstances require.

Responding to overcommitment by counselor-assisted workout is the first level of assistance for a debtor not only because it does not entail reduction of claims but also because participation by all parties is voluntary. The involuntary nature of a creditor's participation in a Chapter 7 or 13 consumer bankruptcy proceeding distinguishes those responses to overextension from that based on counselor-assisted workouts just as certainly as does the fact that these bankruptcy proceedings typically provide something significantly less than full payment of claims. Maintaining counselor-assisted workouts as purely private-sector initiatives serves to continue this distinction and highlights the unique role Consumer Credit Counseling agencies play in the resolution of consumer-debt problems.